I0060848

WORKED EXAMPLES

IN

MASS TRANSFER

(Third Edition)

Benedict Nnolim

First Edition Published August 1993

(c) 1993 B. N. Nnolim

This edition published September 24, 2013
ISBN 978-1-906914-99-8

Other Books by Benedict Nnolim

	Paperback
Fundamentals of Mass Transfer	ISBN 978-1-906914-01-1
Applied Heat Transfer Volume One; Conduction of Heat in Solids (With Worked Examples)	ISBN 978-1-906914-75-2
Applied Heat Transfer Volume Two (With Worked Examples); Heat Convection in Fluids	ISBN 978-1-906914-22-6
The Development and Analysis of New Chemical Plants and Processes	ISBN 978-1-906914-48-6
Worked Examples in Chemical Reaction Engineering	ISBN 978-1-906914-98-1

Ben Nnolim Books
bnbs@ymail.com

PREFACE

Modern academic engineering has become so much more mathematical, predictive and computerised, that students and beginners in engineering subjects, not sufficiently grounded in the fundamentals, are handicapped right from the start.

The approach in this book, therefore, is to attempt to solve this problem in the subject of Mass Transfer, by dealing with, in a question and answer routine, the fundamental concepts, definitions and calculations, of Mass Transfer which are taken for granted in most modern, mainstream, text on the subject.

Many of the problems have been modelled on the sequences/approaches in standard text such as *Mass Transfer*, by R. E. Treybal, *Chemical Engineering* by Coulson and Richardson, *Momentum, Heat and Mass Transfer* by Welty, Wicks and Wilson, and on class problems given by the late Professor Henry Sawistowski of Imperial College, London, to whom I am grateful.

I am grateful to my students over the years whose performance after graduation has justified the validity of this approach, to Miss Augustina Nnamani and to Mrs. Catherine Nwadioke who typed the first and final drafts of this book.

To Our Lady, Queen of Heaven, Saints Jude and Benedict, I am full of praise and thanksgiving and my prayer is that this book and all my life will be to God's purpose.

,
Benedict N. Nnolim
September 9, 2013.

i

Table of Contents

CHAPTER ONE
FUNDAMENTAL CONCEPTS IN MASS TRANSFER

Mass Transfer is one of the four pillars on which chemical engineering operations are anchored.   The others are heat transfer, momentum transfer and chemical reaction/thermodynamics. Although a mass transfer operation will, in a given situation, be closely associated with either heat transfer, momentum transfer, chemical reaction or all of them, the treatment in this book assumes, in the main, that mass transfer occurs at constant heat or momentum transfer and without chemical reaction. Some cases in which mass transfer takes place with heat or momentum transfer or chemical reaction will, however, be discussed.

Industrial processes and operations involve the conversion of materials from one form to another, in a series of steps or processes. These steps or processes, by nature, may be physical, chemical, electrical, electrochemical, mechanical, biological or biochemical, etc. It is preferable that they take place in a known or repeatable manner in order to be useful, industrially.

Changes in mass, during industrial processing of materials, are said to account for more than 50% of the capital investment in most industrial plants (Cussler, 1986) while mass transfer operations account for about 70% of investment costs and 90% of the energy expenses of a typical plant (Buck, 1984). The viability and life of human beings and animals and, between 50 -100% of, the underlying principles behind most major devices and appliances such as automobiles, trains, refrigerators, aeroplanes, etc., depend on mass transfer. It is clear, therefore, that any serious professional in process operations or product manufacturing should be conversant with the principles and practice of mass transfer. Some of these are presented below.

1.1: Mass Transport and Mass Transfer

The need to clarify the meaning of, and the distinctions between, mass transport and mass transfer is a semantic but real one. Standardisation, not only of procedures but also of terminology, is crucial to the economic practice of engineering. Hence, in chemical engineering, it is, generally, accepted that transport

- involves the movement of matter from one place to another within

1

the same phase. Take, for example, any molecule of water, in a stream of water flowing through a pipe. It is said to be moving, in transport, from one place to another within the water phase.

- involves the movement of bulk
- depends on potential differences at the macro level, such as convection, as the driving force.

Transfer, on the other hand

- involves the movement of matter from one phase to another across a phase boundary, known as an interface. Carbon dioxide, for example, is absorbed into lime water across a gas/liquid interface.
- may, especially in multi-component systems, involve differential or opposite movements of parts of the bulk quantity.
- depends on potential differences at the micro-level as the driving force. Such a driving force may be molecular diffusion.

A transfer process, therefore, implies that there exists a mixture, or solution, in which the composition of the component transferred, is such that the component, being transferred, can be defined as the solute.
.
A phase is a collection of homogenous matter within prescribed boundaries. A boundary is a line, plane or surface of discontinuity. Such discontinuities are better determined if the phases and their boundaries are expressed in terms of properties which describe the physical structure or nature of matter.

For example, density, concentration, viscosity, refractive index, chemical potential, fugacity, etc., may be used to describe a given phase such that a difference or change in their value represents a discontinuity or the encounter of a boundary and, hence, another phase. In mass transfer, the boundary between two phases is referred to as an interface.

There are four well known but primitive phases in systems which are encountered in chemical engineering. These are the solid phase, the liquid phase, the vapour phase and the gas phase (such systems are, also, known as P-V-T systems because their equations of state can be expressed in terms of pressure, P, volume, V, and temperature, T). Additional phases may arise in any of these primitive phases as a result of discontinuities in physical property as discussed above. This is, especially, true of solids and liquids where, in any one of solid or liquid phase, there could be several other phases differing in density, viscosity, etc., in equilibrium.

1.2: Mass Transfer Processes and Operations

A mass transfer process is one in which matter is transferred from one phase to another, across an interface, under the influence of a spontaneous driving force such as diffusion. A mass transfer operation, however, is a deliberate process, embarked upon by humans, which makes use of a mass transfer process or processes for economic or other benefit.

What this means is that a mass transfer operation involves both diffusion and a deliberate effort to bring the relevant phases into contact with each other. Such contact can be direct or indirect. Direct contact operations involve the mixing of the phases using any of the known mixing methods such as paddle agitation, gas bubbling, etc. Indirect contact operations involve the use of a medium between the phases. Such a medium may be a solid surface or a membrane.

1.3: Choice of Solvent

Because a mass transfer operation is a deliberate process, the choice of the solvent, in which mass transfer takes place, is, also, deliberate. The choice of solvent is, often, made on the basis of:

i.   Solute solubility and selectivity in the solvent
ii.   Solvent toxicity to humans
iii.   Physical and handling characteristics
iv.   Cost

Solute solubility is the mass of solute which dissolves in a given quantity of solvent under standard temperature and pressure conditions. For liquid and gaseous solutes, it is the volume of solute which dissolves in a given volume of solvent used. Experimentally determined solubility data are preferred to predicted ones and may be found in standard handbooks and data books.

Care should be taken, however, to find out the physical conditions under which these data were obtained. Solubility is a very important consideration in the choice of a solvent since that solvent is to be preferred which dissolves the maximum amount of solute under the given

3

temperature, pressure and operating conditions. If the solute is to be transferred away from a mixture containing other solutes, then selectivity, in addition to solubility, has to be considered. Data on solubility of different solutes in solvents and selectivity of different solvents for solutes can be obtained from chemistry and chemical engineering data handbooks or from predictive thermodynamics using either distribution coefficients or solubility parameters (Snyder, 1979).

Solvent toxicity can arise from either the intrinsic toxicity of the solvent or from the solute - solvent interaction or from both, Toxicity is defined in terms of lethality and carcinogenicity to humans. Both are, usually, determined from tests with animals whose body metabolism is closest to those of humans. In terms of lethality, the Threshold Limit Dose (TLD) is the amount (dose) of the suspected toxic substance which results in the death of 50% of the test animals ingesting the dose. Carcinogens are substances which are capable of inducing a tumour in the test animal or in man. Most countries, and for most industries, have and publish a list of substances, including solvents, which have been found to be toxic or carcinogenic.

Physical and handling characteristics. Highly volatile, flammable, viscous or immiscible solvents have poor handling characteristics in normal industrial use. Some solvents have high heats of vaporisation and are, therefore, expensive to recover after the mass transfer operation because of high energy costs. Such solvents are to be avoided as much as possible.

Cost of Solvent. The cost of a solvent includes:

- The initial purchase price as delivered to the factory site.
- The handling cost, during processing, which is unique to the use of the solvent such as the special precautions that have to be taken.
- The recovery cost from the product by vaporisation, liquid-liquid extraction, decantation, etc.

These costs have to be weighed against the value addition to the product and against the other factors, mentioned above, before the final choice of solvent is made.

1.4: Descriptive Equations

The preferred solution of a technical problem, usually, begins with expressing the problem in mathematical form. One mathematical form, which is as general as can be, is that derived from the first law of thermodynamics for an open system, generally known as either the law of conservation of mass or of energy. When any element of a fluid system, such as that shown in Figure 1, is considered, a mass balance will give a general expression in mathematical form. This general expression relates the intensive and extensive properties of the system to their spatial (space) and temporal (time) coordinates. It is based on the general form of the first law of thermodynamics, for an open system that:

$$Input + Generation/Consumption = Output + Accumulation \qquad (1.1)$$

For the fluid element shown in Figure 1, the expression of equation (1.1), in Cartesian coordinates, is given by Treybal (1984) as

$$\frac{\partial C_A}{\partial t} + u\frac{\partial C_A}{\partial x} + v\frac{\partial C_A}{\partial y} + w\frac{\partial C_A}{\partial z}$$
$$= D_{AB}\left(\frac{\partial^2 C_A}{\partial x^2} + \frac{\partial^2 C_A}{\partial y^2} + \frac{\partial^2 C_A}{\partial z^2}\right) + R_{AV} \qquad (1.2)$$

where u, v, w are the fluid velocities in the x, y and z directions, respectively. C_A is the concentration at any point, in moles per unit volume, R_{AV} its rate of reaction per unit volume and D_{AB} its diffusivity in the mixture consisting of itself A and another component, B.

Figure 1: Mass Balance over a Fluid Element

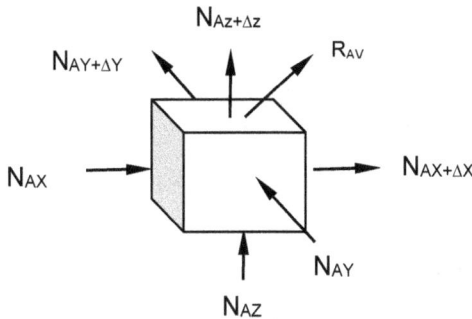

Equation (1.2) is an equation of continuity, in terms of concentration, for a multi-component system (this time, a binary system with A and B as components). Each equation of continuity would be expected to be associated with an equation of motion and an equation of energy. By similar analysis to that of mass, the equations of motion are found to be, in Cartesian co-ordinates, as follows:

In the x – direction

$$\frac{\partial u}{\partial t} + u\frac{\partial u}{\partial x} + v\frac{\partial u}{\partial y} + w\frac{\partial u}{\partial z} = g_x - \frac{1}{\rho}\frac{\partial P}{\partial x} + \frac{\mu}{\rho}\left(\frac{\partial^2 u}{\partial x^2} + \frac{\partial^2 u}{\partial y^2} + \frac{\partial^2 u}{\partial z^2}\right) \quad (1.3)$$

In the y - direction

$$\frac{\partial v}{\partial t} + u\frac{\partial v}{\partial x} + v\frac{\partial v}{\partial y} + w\frac{\partial v}{\partial z} = g_y - \frac{1}{\rho}\frac{\partial P}{\partial y} + \frac{\mu}{\rho}\left(\frac{\partial^2 v}{\partial x^2} + \frac{\partial^2 v}{\partial y^2} + \frac{\partial^2 v}{\partial z^2}\right) \quad (1.4)$$

In the z – direction

$$\frac{\partial w}{\partial t} + u\frac{\partial w}{\partial x} + v\frac{\partial w}{\partial y} + w\frac{\partial w}{\partial z} = g_z - \frac{1}{\rho}\frac{\partial P}{\partial z} + \frac{\mu}{\rho}\left(\frac{\partial^2 w}{\partial x^2} + \frac{\partial^2 w}{\partial y^2} + \frac{\partial^2 w}{\partial z^2}\right) \quad (1.5)$$

The equation of energy is given, in terms of temperature, T, as

$$\frac{\partial T}{\partial t} + u\frac{\partial T}{\partial x} + v\frac{\partial T}{\partial y} + w\frac{\partial T}{\partial z} = \frac{k}{\rho Cp}\cdot\left(\frac{\partial^2 T}{\partial x^2} + \frac{\partial^2 T}{\partial y^2} + \frac{\partial^2 T}{\partial z^2}\right) + \frac{Q_V}{\rho Cp} \quad (1.6)$$

where ρ is the fluid density, P, the absolute pressure, μ, the fluid viscosity, k, the fluid thermal conductivity, Cp, the heat capacity and Q_V, the rate of heat generation per unit volume by chemical reaction. The same equations are given, in cylindrical and spherical co-ordinates, in the Appendix.

The majority of solutions to problems in diffusion in mass transfer are based on defining the appropriate initial and boundary conditions which are, then, applied to the above equations. For example, for isothermal and molecular diffusion in a stagnant medium and at constant pressure, the following simplifications become possible.

In a stagnant medium,
$$u = v = w = 0 \quad (1.7)$$

At constant pressure;
$$\frac{\partial P}{\partial x} = \frac{\partial P}{\partial y} = \frac{\partial P}{\partial z} = 0 \quad (1.8)$$

At constant temperature

$$\frac{\partial T}{\partial x} = \frac{\partial T}{\partial y} = \frac{\partial T}{\partial z} = 0 \qquad (1.9)$$

When these are substituted in the equations of continuity, motion and energy, it will be found that only the portion of the continuity equation (1.2), written below, will remain. That is

$$\frac{\partial C_A}{\partial t} = D_{AB} \left(\frac{\partial^2 C_A}{\partial x^2} + \frac{\partial^2 C_A}{\partial y^2} + \frac{\partial^2 C_A}{\partial z^2} \right) + R_{AV} \qquad (1.10)$$

Since there is no chemical reaction, $R_{AV} = 0$. For unidirectional diffusion, in which there are no variations in C_A along the y and z directions, molecular diffusion is in the x-direction only. Thus

$$\frac{\partial C_A}{\partial t} = D_{AB} \frac{\partial^2 C_A}{\partial x^2} \qquad (1.11)$$

If steady state conditions prevail,

$$\frac{\partial C_A}{\partial t} = 0 \qquad (1.12)$$

and

$$D_{AB} \frac{\partial^2 C_A}{\partial x^2} = 0 \quad or \quad D_{AB} \frac{d C_A}{d x} = constant \qquad (1.13)$$

From Fick's first law of molecular diffusion, this constant is equal to the molar flux, N_A, so that

$$N_A = - D_{AB} \frac{d C_A}{d x} \qquad (1.14)$$

with the boundary conditions that at

$$x = 0, \quad C_A = C_{A0} \text{ and at } x = L, \quad C_A = 0 \qquad (1.15)$$

If unsteady state conditions prevail, then equation (1.11) prevails with the initial and boundary conditions

t = 0	x = 0	$C_A = C_{A0}$		
	x = L	$C_A = 0$		1.16a
t > 0	x = 0	$C_A = C_{A0}$		
	x = L	$C_A = C_{A0}$		1.16b

Equation (1.11) and the initial and boundary conditions described by (1.16a) and (1.16b) provide a mathematical description of unsteady state molecular diffusion into a semi-infinite medium.

When chemical reaction occurs, the relevant simplification, for one dimensional diffusion, is

$$\frac{\partial C_A}{\partial t} = D_{AB}\frac{\partial^2 C_A}{\partial x^2} + R_{AV} \tag{1.17}$$

Another typical example is that of molecular diffusion in steady state, isothermal fluid flow, over a flat plate and with chemical reaction. For changes occurring only in the x and y directions, i.e., two dimensional flow, the equations of motion, energy and continuity reduce to (Bird, Stewart and Lightfoot, 1960)

$$\frac{\partial u}{\partial x} + \frac{\partial v}{\partial y} = 0 \tag{1.18}$$

for continuity of mass and

$$u\frac{\partial u}{\partial x} + v\frac{\partial u}{\partial y} = \frac{\mu}{\rho}\frac{\partial^2 u}{\partial y^2} \tag{1.19}$$

for the equation of motion. The equation for continuity of concentration of A is given by

$$u\frac{\partial C_A}{\partial x} + v\frac{\partial C_A}{\partial y} = D_{AB}\frac{\partial^2 C_A}{\partial y^2} + R_{AV} \tag{1.20}$$

Yet another example involves, not only steady state mass transfer but also, simultaneous heat and momentum transfer as well and for flow over a flat plate. In such a situation, without chemical reaction, the equations are the same as equations (1.18) and (1.19) for continuity of mass and motion respectively. The equation for continuity of concentration of A is

$$u\frac{\partial C_A}{\partial x} + v\frac{\partial C_A}{\partial y} = D_{AB}\frac{\partial^2 C_A}{\partial y^2} \tag{1.21}$$

while that for energy is

$$u\frac{\partial T}{\partial x} + v\frac{\partial T}{\partial y} = \frac{k}{\rho Cp}\frac{\partial^2 T}{\partial y^2} \tag{1.22}$$

It can be seen, from the above, that many conditions are possible depending on the initial and boundary conditions of the system under consideration. When analysing mass transfer in cylindrical enclosures, conduits or objects, the polar co-ordinate system is recommended. When spherical enclosures, conduits or objects are involved, the spherical co-ordinate system is recommended.

This summary is, only, intended to provide perspective to the student reading this book. The mass transfer situations, actually, treated in this book, revolve around systems in which either

$$D_{AB} \frac{dC_A}{dx} = constant \qquad \textit{see equation (1.13)}$$

or

$$\frac{\partial C_A}{\partial t} = D_{AB} \frac{\partial^2 C_A}{\partial x^2} \qquad \textit{see equation (1.11)}$$

Such systems, though simple looking in mathematical form, provide the basis for the solution of a very large number of practical problems in commercial mass transfer processes and operations. The student would do well, therefore, to master them.

References for Chapter One

1   Bird R. B., Stewart W. E., and Lightfoot E. N.; *Transport Phenomena*, John Wiley & Sons Inc., New York, USA. 1960

2   Buck E, (1984), *Applying Phase Equilibrium Thermodynamics*, CHEMTECH, 14, No. 8, pp 570 - 575, American Chemical Society, Washington D.C., USA.

3   Cussler E. L (1986), *A Mass Transfer Tutorial*, CHEMTECH, 16, No. 7, pp 422 - 425, American Chemical Society, Washington D.C.,USA

4   Treybal R. E., *Mass Transfer Operations*, McGraw -Hill Book Company, New York, 1980

5   Snyder L., (1979), *Solution to Solution Problems - 1*, CHEMTECH, **9**, No. 12, pp 750 -754, American Chemical Society, Washington D.C., USA.

CHAPTER TWO
MOLECULAR DIFFUSION

This chapter presents eight problems and their solutions in three parts. The intention of the first part is to enable the student appreciate the meaning of molar flux, molar average velocity and what happens in equi-molar counter-diffusion and in diffusion through a stagnant medium, under isothermal conditions in which there are no chemical reactions or bulk velocity gradients (Examples 2.1 to 2.5).

The second set of problems illustrates how molecular diffusivity may be estimated using different types of correlations, derived from empirical observation or a combination of theoretical physical chemistry and empirical measurements (Examples 2.6 to 2.12).

The third part gives a summary of current theories of mass transfer across an interface. Only clean and stable interfaces are assumed.

Example 2.1

Define the molar average velocity and the molar flux in mass transfer for the situations in which

        a)   there is no bulk motion
        b)   there is bulk motion.

If the molar masses of NaCl and water are 58.5 and 18, respectively, and they move with velocities of 15 and 11.4 cm/s, respectively, in a 10% NaCl solution in water, determine their respective molar fluxes relative to their molar average velocity. The density of the salt solution is given as $1071 kg/m^3$

Answer

The molar average velocity, U, is, generally, defined as the molar concentration weighted average of all velocities of all the constituents in the mixture. That is

$$\overline{U} = \frac{\sum\limits_{i=1}^{n} \dfrac{\rho_i}{M_i} u_i}{\sum\limits_{i=1}^{n} \dfrac{\rho_i}{M_i}} \tag{1}$$

11

where ρ_i = density of component, i
$\quad$ M_i = molar mass of component, i.
$\quad$ u_i = velocity of component, i

The molar flux, N_i, is the moles flowing, per unit time, per unit area, normal to the direction of flow. That is

$$N_i = \frac{\rho_i}{M_i}(u_i - u_{ref}) \qquad (2)$$

where u_{ref} = reference velocity relative to which motion is measured. When there is no bulk motion, $u_{ref} = 0$ and

$$N_i = \frac{\rho_i}{M_i}u_i \qquad (3)$$

When there is bulk flow, a reasonable reference velocity is $u_{ref} = u_1$ Then

$$N_i = \frac{\rho_i}{M_i}(u_i - u_1) \qquad (4)$$

Density of the 10% solution = density contribution of NaCl + density contribution of water. Since density contribution = actual density x mass fraction,

$$Density\ contribution\ of\ water = 1000\ x\ 0.9 = 900\ kg\,/\,m^3 \quad (5)$$

$$Density\ contribution\ of\ NaCl = 1071 - 900 = 171\ kg\,/\,m^3 \quad (6)$$

Thus

$$Actual\ density\ of\ NaCl = \frac{171}{0.1} = 1710\ kg\,/\,m^3 \qquad (7)$$

Hence

$$Molal\ Average\ Velocity,\ \overline{U} = \frac{\dfrac{1710}{58.5}x15 + \dfrac{1000}{18}x11.4}{\dfrac{1710}{58.5} + \dfrac{1000}{18}}$$

$$= \frac{438.46 + 633.33}{29.23 + 55.56} = 12.64\ m\,/\,s \qquad (8)$$

Molar Flux for NaCl

$$N_{NaCl} = \frac{1710}{58.5}(15 - 12.64)x\frac{1}{100}\frac{kg}{m^3}\frac{kmol}{kg}\frac{cm}{s}\frac{m}{cm} = 0.69\frac{kmol}{m^2 s} \quad (9)$$

Molar Flux for H$_2$O

$$N_{H_2O} = \frac{1000}{18}(11.4 - 12.64)x\frac{1}{100}\frac{kg}{m^3}\frac{kmol}{kg}\frac{cm}{s}\frac{m}{cm} = -0.69\frac{kmol}{m^2 s} \quad (10)$$

Thus, NaCl and H$_2$O are in equimolar counter-diffusion (EMCD).

Example 2.2

Fick's first law for molecular diffusion in a binary system of components, A and B. gives the molar flux of A as

$$n_A = - D_{AB} \frac{\partial C_A}{\partial z} \qquad (a)$$

while the molar flux of A, for diffusion in a binary system, is, also, given as

$$N_A = X_A \left(N_A + N_B \right) - D_{AB} \frac{\partial C_A}{\partial z} \qquad (b)$$

where

N_A, N_B = molar flux of A and B, respectively
X_A = mole fraction of A in the system
D_{AB} = molecular diffusivity of A in B
$\dfrac{\partial C_A}{\partial z}$ = concentration gradient of A along the z - direction.

Explain each equation and the situation to which it applies.

Answer

The Fick's first law equation is an expression, in a mass transfer situation, of the general relationship between transport, or transfer, of a component A per unit area, the driving force for the transport, or transfer and the resistance involved. For such a situation:

$$Transfer\ or\ Transport\ per\ unit\ area = \frac{Driving\ Force}{Re\,sis\,tan\,ce} \qquad (1)$$

Since diffusion is a transport process, let us examine the molar flux, N_A, of component, A, in a binary system of A and B. This molar flux, is the transfer or transport per unit area. Notice that because it is a flux, by definition, it has, already, units of moles per unit area, such as moles /m². The driving force for the diffusion of A in the system is, of course, the concentration difference of A between the regions in which diffusion is occurring. That is

$$Driving\ Force\ for\ A = Concentration\ Difference = d\,C_A, \frac{mol}{m^3} \qquad (2)$$

The conductance for A is

$$Conduc\tan ce\ of\ A = \frac{moles}{unit\ concentration\ difference} = \frac{n_A}{\left(\dfrac{\partial C_A}{\partial z} \right)} = D_{AB} \qquad (3)$$

13

Resistance is the inverse of the conductance. That is

$$Re\,sis\,\tan ce\ to\ A = \frac{1}{\dfrac{D_{AB}}{d\,z}} = \frac{\left(\dfrac{\partial C_A}{\partial z}\right)}{n_A} = \frac{1}{D_{AB}} \qquad (4)$$

Combining these various definitions, we get that

$$n_A = \frac{d\,C_A}{\left(\dfrac{d\,z}{D_{AB}}\right)} \qquad (5)$$

But n_A, D_{AB} and dz are all positive while dC_A is negative. To get the signs correct, a negative sign is introduced to make dC_A positive. Because there are two components, partial derivatives are used in dC_A and dz to distinguish changes due solely to A and those due solely to B. The result is the Fick's first law equation for molecular diffusion. It shows, also, that neither bulk flow nor convective motion, but only molecular diffusion, is involved in the Fick's first law equation.

The second equation (equation (b) above) applies to a situation in which there is bulk or convective motion in addition to molecular diffusion. Thus since

Molar Flux of A = Convective molar flux + diffusiond molar flux

and

Total Molar Flux = molar flux of A + molar flux of B

That is

$$N_T = N_A + N_B \qquad (6)$$

then *Convective Molar Flux of $A = X_A N_T = X_A(N_A + N_B)$* $\quad(7)$

But *Diffusional molar flux,* $n_A = -D_{AB}\dfrac{\partial C_A}{\partial z}$ *from* (a)

Hence the total molar flux of A is then

$$N_A = X_A(N_A + N_B) - D_{AB}\frac{\partial C_A}{\partial z} \qquad (8)$$

Note, also, that the total molar flux of B is

$$N_B = X_B(N_A + N_B) - D_{BA}\frac{\partial C_B}{\partial z} \qquad (9)$$

Example 2.3

The molar flux, for steady state diffusion of components, A and B, across an interface, in a binary system in laminar flow, is given by

$$N_A = U_Z C_T + n_A \tag{1}$$

$$N_B = U_Z C_T + n_B \tag{2}$$

where N_A and N_B are the molar flux of A and B, respectively, U_Z is the molar average velocity due to bulk motion, C_T is the total concentration of components, A and B in the system, n_A, n_B the molar flux of A and B, respectively, due to molecular diffusion only and

$$n_A = -D_{AB} \frac{\partial C_A}{\partial z} \tag{3}$$

$$n_B = -D_{BA} \frac{\partial C_B}{\partial z} \tag{4}$$

D_{AB} and D_{BA} are the diffusion coefficients of A in B and B in A, respectively, z is the distance from the interface in the direction of diffusion. Show that, for a gaseous system for which $PV = nRT$,

$$N_A = \frac{N_A}{N_A + N_B} \cdot \frac{P_T}{R.T} \cdot \frac{D_{AB}}{z} \cdot \ln\left[\frac{N_A/(N_A + N_B) - P_{A2}/P_T}{N_A/(N_A + N_B) - P_{A1}/P_T}\right] \tag{5}$$

where P_T = total system pressure, P_{A1}, P_{A2} = the partial pressures in phases 1 and 2, respectively, on either side of the interface. How would this expression be affected if A and B were in equimolar counter-diffusion?

Answer

Since

$$U_Z C_T = X_A(N_A + N_B) = \frac{C_A}{C_T}(N_A + N_B) \tag{6}$$

and

$$n_A = -D_{AB} \frac{\partial C_A}{\partial z} \qquad \textit{from} \quad (3)$$

$$n_B = -D_{BA} \frac{\partial C_B}{\partial z} \qquad \textit{from} \quad (4)$$

Then

$$N_A = \frac{C_A}{C_T}(N_A + N_B) - D_{AB} \frac{\partial C_A}{\partial z} \tag{7}$$

which, on re-arrangement for integration, gives

$$\int_{C_{A1}}^{C_{A2}} \frac{dC_A}{N_A C_T - C_A(N_A + N_B)} = -\int_0^z \frac{dz}{D_{AB} C_T} \tag{8}$$

When N_A, N_B, D_{AB}, and C_T are constant, the integration yields

$$\frac{1}{(N_A + N_B)} \ln \left| N_A C_T - C_A (N_A + N_B) \right|_{C_{A1}}^{C_{A2}} = \frac{z}{D_{AB} C_T} \quad (9)$$

On further simplification and rearrangement, we get

$$\frac{1}{(N_A + N_B)} \ln \left[\frac{N_A C_T - C_{A2}(N_A + N_B)}{N_A C_T - C_{A1}(N_A + N_B)} \right] = \frac{z}{D_{AB} C_T} \quad (10)$$

Dividing the terms within the logarithm term by $(N_A + N_B)/C_T$ and multiplying both sides of the equation by N_A, we get that

$$\frac{N_A}{N_A + N_B} . \ln \left[\frac{N_A/(N_A + N_B) - C_{A2}/P_T}{N_A/(N_A + N_B) - C_{A1}/P_T} \right] = \frac{N_A . z}{D_{AB} . C_T} \quad (11)$$

Since $PV = nRT$ and $C = n/V = P/RT$ then $C_T = P_T/RT$, $C_{A1} = P_{A1}/RT$, and $C_{A2} = P_{A2}/RT$. If the system is isothermal, $T_1 = T_2 = T = $ constant so that $C_{A2}/C_T = P_{A2}/P_T$, etc.

Then
$$\frac{N_A . z}{D_{AB}} . \frac{RT}{P_T} = \frac{N_A}{N_A + N_B} . \ln \left[\frac{N_A/(N_A + N_B) - P_{A2}/P_T}{N_A/(N_A + N_B) - P_{A1}/P_T} \right] \quad (12)$$

Thus

$$N_A = \frac{N_A}{N_A + N_B} . \frac{P_T}{R.T} . \frac{D_{AB}}{z} . \ln \left[\frac{N_A/(N_A + N_B) - P_{A2}/P_T}{N_A/(N_A + N_B) - P_{A1}/P_T} \right] \quad Ans$$

For equi-molar counter-diffusion, $N_A + N_B = 0$ and the above equation cannot be used. The equation to use is

$$N_A = \frac{C_A}{C_T}(N_A + N_B) - D_{AB} \frac{dC_A}{dz} \quad from \quad (7)$$

which, since $N_A + N_B = 0$, reduces to

$$N_A = -D_{AB} \frac{dC_A}{dz} = -\frac{D_{AB}}{RT} \frac{dP_A}{dz} \quad (13)$$

since $C_A = P_A/RT$. That is, the effect is as if there was no bulk motion. Thus

$$N_A \int_0^z dz = -\frac{D_{AB}}{RT} \int_{P_{A1}}^{P_{A2}} dP_A \quad from \ which \quad N_A = \frac{D_{AB}}{zRT}(P_{A2} - P_{A1}) \quad Ans$$

Example 2.4

Oxygen is diffusing in an oxygen - nitrogen mixture at 1 atm. and 25 C, across planes 2 mm apart. The concentration of oxygen in these planes is 10% and 20%, by volume, respectively. Calculate the molar flux of oxygen for equi-molar counter-diffusion. Take oxygen to be component A, nitrogen to be component B and $D_{AB} = 2.11 \times 10^{-5}$, m^2/s (Treybal,

1980). The universal gas constant, R = 8.314 kJ/kmol.K and 1 atm = 1.013×10^5, N/m².

Answer

Molar flux of oxygen,

$N_A = $ *flux due to pure diffusion + flux due to bulk motion.*

That is

$$N_A = -D_{AB} \frac{d C_A}{d z} + X_A \left(N_A + N_B \right) \qquad (1)$$

where N_B = molar flux of nitrogen. For equi-molar counter-diffusion, $N_A = -N_B$ or $N_A + N_B = 0$ so that

$$N_A = -D_{AB} \frac{d C_A}{d z} \qquad (2)$$

Considering oxygen as an ideal gas, $C_A = p_A/RT$ so that

$$N_A = -\frac{D_{AB}}{RT} \frac{d P_A}{d z} \qquad (3)$$

which, on integration and re-arrangement, yields

$$N_A = \frac{D_{AB}}{z RT} \left(P_{A2} - P_{A1} \right) \qquad (4)$$

For gas mixtures,

Partial pressure = mole fraction x total pressure.
Mole fraction = volume fraction

Hence, from the data given,

$P_{A1} = 0.2 P_T$ $D_{AB} = 2.\,11 \times 10^{-5}$,m²/s
$P_{A2} = 0.\,1 P_T$ $z = 0.00$ m
$T = 298$ K $R = 8.314$ kN.m/kmol.K

Substituting these in equation (4)

$$N_A = \frac{2.11 \times 10^{-5}}{0.002 \times 8.314 \times 10^3 \times 298} (0.2 - 0.1) \times 1.013 \times 10^5,$$

$$\frac{m^2}{s} \frac{1}{m} \frac{kmol.K}{N.m} \frac{1}{K} \frac{N}{m^2}$$

17

$$= 4.314 x 10^{-5} \frac{kmol}{m^2 .s} \quad Ans$$

Example 2.5

Starting from the molar flux equation for laminar flow,

$$N_A = -D_{AB} \frac{d C_A}{d z} + X_A (N_A + N_B) \quad (a)$$

show that for the diffusion of component, A, through stagnant, non-diffusing component, B, of thickness, z, its molar flux, N_A, is given by

$$N_A = -\frac{D_{AB}}{z} . \frac{1}{X_{BM}} . \left(\frac{\rho}{M}\right)_{avg} (X_{A1} - X_{A2}) \quad (b)$$

where N_A = molar flux of A, kmol/m^2.s; D_{AB} = molecular diffusivity of A in B, m^2/s; X_{A1}, X_{A2} - mole fractions of A at points 1 and 2, z meters apart; $(\rho/M)_{avg}$ = average molar density of A in the medium; X_{BM} = log mean mole fraction of the non-diffusing component such that

$$X_{BM} = \frac{X_{B2} - X_{B1}}{\ln(X_{B2}/X_{B1})} \quad (c)$$

Answer

Note that this is an extension of Example 2.3 applied to the liquid phase. Given that

$$N_A = -D_{AB} \frac{d C_A}{d z} + X_A (N_A + N_B) \quad (1)$$

and $X_A = C_A/C_T$, then

$$N_A = -D_{AB} \frac{d C_A}{d z} + \frac{C_A}{C_T} (N_A + N_B) \quad (2)$$

Since N_A, N_B, D_{AB}, and C_T are constant,

$$\int_0^z dz = -D_{AB} C_T \int_{C_{A1}}^{C_{A2}} \frac{d C_A}{N_A C_T - C_A (N_A + N_B)} \quad (3)$$

so that

$$z = \frac{D_{AB} C_T}{(N_A + N_B)} . \ln\left[\frac{N_A/(N_A + N_B) - C_{A2}/C_T}{N_A/(N_A + N_B) - C_{A1}/C_T}\right] \quad (4)$$

And, multiplying both sides by N_A and rearranging,

$$N_A = \frac{N_A}{(N_A + N_B)} . \frac{D_{AB} C_T}{z} . \ln\left[\frac{N_A/(N_A + N_B) - C_{A2}/C_T}{N_A/(N_A + N_B) - C_{A1}/C_T}\right] \quad (5)$$

Since B is not diffusing $N_B = 0$. Also $X_{A2} = C_{A2}/C_T$, $X_{A1} = C_{A1}/C_T$ and $C_T = (\rho/M)_{avg}$. Hence

$$N_A = \frac{D_{AB}}{z}\cdot\left(\frac{\rho}{M}\right)_{avg}\cdot\ln\left[\frac{1 - X_{A2}}{1 - X_{A1}}\right] = \frac{D_{AB}}{z}\cdot\left(\frac{\rho}{M}\right)_{avg}\cdot\ln\left[\frac{X_{B2}}{X_{B1}}\right] \quad (6)$$

since $X_B = 1 - X_A$ for a binary system and

$$X_{BM} = \frac{X_{B2} - X_{B1}}{\ln\left(X_{B2}/X_{B1}\right)} \qquad from \quad (c)$$

Thus

$$\ln\frac{X_{B2}}{X_{B1}} = \frac{X_{B2} - X_{B1}}{X_{BM}} = \frac{(1 - X_{A2}) - (1 - X_{A1})}{X_{BM}} = \frac{X_{A1} - X_{A2}}{X_{BM}} \quad (7)$$

Substituting (7) in (6)

$$N_A = \frac{D_{AB}}{z}\cdot\left(\frac{\rho}{M}\right)_{avg}\cdot\frac{X_{A1} - X_{A2}}{X_{BM}} \qquad Ans.$$

Example 2.6

The diffusivity of a vapour in a gas is given by the equation due to Fuller *et al* (1966),

$$D_V = \frac{1.013 \times 10^{-7} T^{1.75}\left(\frac{1}{M_A} + \frac{1}{M_B}\right)^{\frac{1}{2}}}{P\left[\left(\sum_A V_i\right)^{\frac{1}{3}} + \left(\sum_B V_i\right)^{\frac{1}{3}}\right]^2}$$

where D_v = diffusivity of vapour, m²/s; T = temperature of the system, deg. K; M_A, M_B = molar mass of component A and B, respectively; P = total pressure, bar and $\sum_A V_i$, $\sum_B V_i$ the summation of the specific *diffusion volume coefficients* for components A and B, respectively. Estimate the diffusivity of water vapour in air at 26 C, 1 atm., given that $\sum_A V_i = 20$ for air and $\sum_B V_i = 9.44$ for water vapour. Take 1 atm = 1.013 bar.

Answer

This problem tests your ability to substitute values in, and to evaluate, formulae and correlations correctly.

Since T = 26 + 273 = 299 K; P = 1 atm. = 1.013 bar; M_A = 18 for water (the diffusing solute); M_B = 28.84 for air (the non-diffusing solvent)

$$D_V = \frac{1.013x10^{-7}x(299)^{1.75}\left(\frac{1}{18}+\frac{1}{28.84}\right)^{\frac{1}{2}}}{1.013x\left[(20)^{\frac{1}{3}}+(9.44)^{\frac{1}{3}}\right]^2} = 2.77x10^{-5}, \frac{m^2}{s} \quad Ans$$

Example 2.7

According to Wilke and Chang (1955), liquid diffusivity is given by

$$D_L = \frac{1.173x10^{-16}(\Phi_L.M)^{0.5}T}{\mu.V_M^{0.6}}$$

where D_L = liquid diffusivity, m^2/s; Φ_L = association factor for the solvent; M = molar mass for the solvent; μ = viscosity of the solvent, Ns/m^2; T = temperature of the system, K; V_M = molar volume of the solvent at its boiling point, $m^3/kmol$. Estimate the diffusivity of ethanol in water at 30 C given that Φ_L = 2.26 for water, μ = 0.86 mNs/m^2 and V_M = $\Sigma(n_iV_i)$; where n_i = number of atoms in a solute molecule, V_i = atomic volume of atom i in the solute and the atomic volumes are given below:

Atom	Atomic Volume, V_i, $m^3/kmol$
C	0.0148
H	0.0037
O	0.0074

Answer

Since ethanol has the formula C_2H_5OH, i.e. (C_2H_6O), the molar volume is given by

V_M =2 x 0.0148 + 6 x 0.0037 + 1 x 0074 = 0.0592, $m^3/kmol$.
T = 30 + 273 = 303K.
0.86 mNs/m^2 = 0.001 x 0.86 Ns/m^2

so that

$$D_L = \frac{1.173x10^{-16}(2.26x18)^{0.5}x303}{0.001x0.86x(0.0592)^{0.6}} = 1.437x10^{-9}, \frac{m^2}{s} \quad Ans$$

Example 2.8

It is desired to estimate the diffusivity of n-propane, A, in nitrogen, B, at 30 C and 60 C at 1 atm. The Hirschfelder, Bird and Spotz equation (Treybal, 1980), given below, and recommended for mixtures of non-polar or polar with non-polar gases, is to be used.

$$D_{AB} = \frac{10^{-4}\left[1.084 - 0.249\left(\frac{1}{M_A} + \frac{1}{M_B}\right)^{\frac{1}{2}}\right]T^{1.5}\left(\frac{1}{M_A} + \frac{1}{M_B}\right)^{\frac{1}{2}}}{P_T \cdot \sigma_{AB}^2 \cdot f\left(\frac{kT}{\varepsilon_{AB}}\right)}$$

where D_{AB} = diffusivity, m^2/s; T = absolute temperature, K; M_A, M_B = molecular weight of A and B, respectively, kg/kmol; P_T = absolute pressure, N/m^2; σ_{AB} = molecular separation at collision, $nm = (\sigma_A + \sigma_B)/2$; ε_{AB} = energy of molecular attraction = $\sqrt{(\varepsilon_A \varepsilon_B)}$; k = Boltzman's constant; $f(kT/\varepsilon_{AB})$ = collision integral given as a function of kT/ε_{AB} as shown below.

kT/ε_{AB}	$f(kT/\varepsilon_{AB})$
1.0	0.72
1.5	0.60
2.0	0.54
2.8	0.48
3.0	0.47
4.0	0.44

You are given that, for nitrogen, ε/k = 71.4 K; σ = 0.3798 nm and for n-propane, , ε/k =237.1 K; σ_{AB} = 0.5118 nm. Also the atomic weights are N = 14; C = 12; H = 1 and 1 atm. = 1.013 x 10^5, N/m^2.

Answer

The molecular weight of n-propane, C_3H_8, is 3 x 12 + 1 x 8 = 44. That of nitrogen is 2 x 14 = 28. Hence

$$\left(\frac{1}{M_A} + \frac{1}{M_B}\right)^{\frac{1}{2}} = \left(\frac{1}{44} + \frac{1}{28}\right)^{\frac{1}{2}} = 0.242 \qquad (2)$$

$$\sigma_{AB} = \frac{\sigma_A + \sigma_B}{2} = \frac{0.5118 + 0.3798}{2} = 0.446 \qquad (3)$$

21

$$\frac{\varepsilon_{AB}}{k} = \sqrt{\frac{\varepsilon_A}{k} \cdot \frac{\varepsilon_B}{k}} = \sqrt{237.1 \, x \, 71.4} = 130.1 \qquad (4)$$

At 30 C, T = 30 + 273 = 303 K.

$$\left(\frac{kT}{\varepsilon}\right)_{AB} = \frac{303}{130.1} = 2.329 \qquad (5)$$

$f(kT/\varepsilon_{AB})$ is estimated by interpolation using the values in the Table above as

$$\frac{f\left(\frac{kT}{\varepsilon}\right) - 0.54}{0.48 - 0.54} = \frac{2.329 - 2.0}{2.8 - 2.0} = 0.411 \quad or \quad f\left(\frac{kT}{\varepsilon}\right) = 0.515 \qquad (6)$$

Hence

$$D_{AB} = \frac{10^{-4} \left[1.084 - 0.249 \, x \, 0.242\right] x \, (303)^{1.5} \, x \, 0.242}{1.013 \, x \, 10^5 \, x \, (0.446)^2 \, x \, 0.515} = 1.259 \, x \, 10^{-5} \quad Ans$$

At 60 C, T = 60 + 273 = 333 K.

$$\left(\frac{kT}{\varepsilon}\right)_{AB} = \frac{333}{130.1} = 2.560 \qquad (7)$$

$f(kT/\varepsilon_{AB})$ is estimated by interpolation using the values in the Table above as

$$\frac{f\left(\frac{kT}{\varepsilon}\right) - 0.54}{0.48 - 0.54} = \frac{2.560 - 2.0}{2.8 - 2.0} = 0.700 \quad or \quad f\left(\frac{kT}{\varepsilon}\right) = 0.498 \qquad (8)$$

Hence

$$D_{AB} = \frac{10^{-4} \left[1.084 - 0.249 \, x \, 0.242\right] x \, (333)^{1.5} \, x \, 0.242}{1.013 \, x \, 10^5 \, x \, (0.446)^2 \, x \, 0.498} = 1.500 \, x \, 10^{-5} \, Ans$$

Example 2.9

The mean diffusion coefficient, D_{AM}, of component A in a multi-component mixture consisting of components A, B, C, D, etc., is given by

$$D_{AM} = \frac{1}{\frac{y_B}{D_{AB}} + \frac{y_C}{D_{AC}} + \frac{y_D}{D_{AD}} + \ldots \ldots} \qquad (1)$$

where D_{AB}, D_{AC}, D_{AD}, etc., are the binary diffusion coefficients between A and each of the constituents, B, C, D, etc., and y_A, y_B, y_C, y_D, etc., are the mole fractions of A, B, C, D, etc., in the multi-component mixture. Calculate the mean diffusion coefficient of A in a gas mixture at 5 atm., and 25.9 C, described as follows:

Component	Name	Partial Pressure, atm	Binary Mixture	Binary Diffusion Coefficient, m²/s
A	Air	2.0		
B	Water Vapour	1.5	Air-Water Vapour	2.58 x 10⁻⁵
C	Toluene	0.5	Air-Toluene	0.86 x 10⁻⁵
D	n-Butanol	1.0	Air-n-Butanol	0.87 x 10⁻⁵

Answer

The mole fractions of the components in the mixture are estimated as the ratios of the partial to the total pressures as follows

y_B	y_C	y_D
1.5/5 = 0.3	0.5/5 = 0.1	1.0/5 = 0.2

When these are substituted onto the equation for diffusivity, we get

$$D_{AM} = \frac{1}{\dfrac{0.3}{2.58 x 10^{-5}} + \dfrac{0.1}{0.86 x 10^{-5}} + \dfrac{0.2}{0.87 x 10^{-5}}} = 2.162 x 10^{-5} \frac{m^2}{s}$$

Example 2.10

Discuss, briefly, the various theories of mass transfer across an interface. Include, in your discussions, models of surface renewal and how the theories and models estimate the mass transfer coefficient.

Answer

The theories can be classified into those that involve molecular diffusion only, eddy diffusion only or both. Models of surface renewal, on the other hand, are described using surface age distribution functions. One class of these functions predicts regular surface renewal, another, random surface renewal while another models a surface as a series of stirred tanks. Combining the mass transfer theory with a surface renewal model permits greater flexibility in dealing with a lot of real life mass transfer problems.

Theories Incorporating Molecular Diffusion

The Whitman Film Theory
(Whitman W. G. (1923); Chem, Met. Eng.; 29; 147)

This is a steady state model which defines an equivalent film thickness, in each phase, in which all resistance to mass transfer, in that phase, resides. In the gas phase, for example, the concentration versus distance profile, starting from the gas interface, looks like this

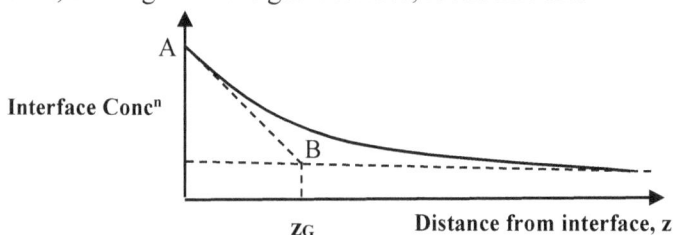

The model assumes that the curve equivalent to the actual concentration curve is the straight line AB and hence that the mass transfer coefficient is directly proportional to the molecular diffusivity. That is

$$Mass\ transfer\ coefficient,\ k_M = \frac{Molecular\ Diffusivity}{Equivalent\ Film\ Thickness} \propto D^{1.0}\ (1)$$

The equivalent film thickness is obtained at the intersection of line AB with the asymptote along the z-axis to the actual concentration curve. When compared to other theories of mass transfer at an interface, it is seen to be a steady state model at very long times of contact.

The Penetration Theory
{Higbie R. (1935): Trans. Am. Inst. Chem. Engrs; 31, 365}

This is an unsteady state model which describes mass transfer at very short times and in the region near the interface. This region is regarded as semi-infinite because, although interfacial concentration is, instantaneously, established, the significant depth of penetration of solute, into the phase, is smaller than the depth of undisturbed phase. The model estimates that the concentration at any point in the semi-infinite medium, away from the interface, depends on contact time.

The plot below illustrates the concentration curve for increasing contact times. The model, also, estimates that the mass transfer coefficient is directly proportional to the square root of molecular diffusivity. That is

$$Mass\ transfer\ coefficient,\ k_M = \sqrt{\frac{Molecular\ \ Diffusivity}{\pi\ x\ Contact\ Time}} \propto D^{\frac{1}{2}}\ (2)$$

Schematic of the Higbie Theory

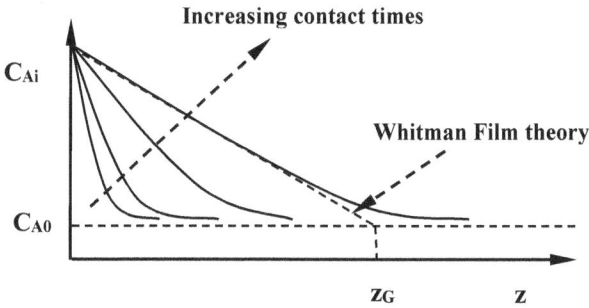

The Film Penetration Theory
{*Toor H. L.and Marchello J.M. (1958); AIChE, 4; 97*}

This theory incorporates both the steady state and unsteady state notions of the Whitman film and the Higbie's penetration, theories. It postulates that the Whitman film theory represents the steady state limit in mass transfer for very long contact times while the penetration theory describes the unsteady state limit for very short contact times. Consequently, the concentration profile of this theory approximates that of the penetration theory at very short contact times and that of the Whitman film theory at very long contact times with intermediate values in between.

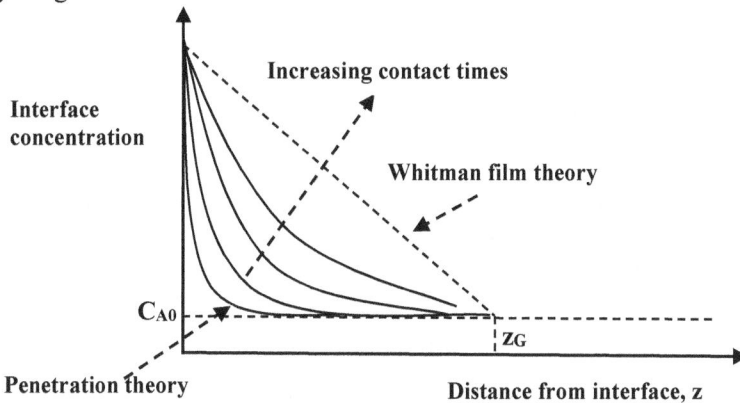

This theory estimates the mass transfer coefficient from either the Whitman or Penetration theory depending on the regime applicable.

Models Of Surface Renewal

While the mass transfer theories, above, attempt to predict the instantaneous transfer of mass on a molecular level, they do not reflect the total or average value on a macro scale over the time of contact. It is the surface age distribution function which allows us to average or total the micro scale transfer to reflect its value on a macro scale. Thus, if the instantaneous molar flux of component A, at the interface, is N_{Ai} (θ) and the surface age distribution function is $\Phi(\theta)$, where θ_C is time of contact, then the total molar flux, N_{Ai}, on a macro scale, is

$$N_{Ai} = \int_0^{\theta_C} N_{Ai}(\theta)\Phi(\theta)d\theta \qquad (3)$$

The Higbie or Regular Surface Renewal Model
(Higbie R. (1935): Trans. Am. Inst. Chem. Engrs; 31, 365)

This model postulates that all surface elements have the same chance of exposure at the interface. Mass transfer, however, takes place only for surface elements exposed for a fixed time of contact, θ_C, which is the same for all elements. The surface age distribution function, $\Phi(t) = 1/\theta_C$, is a step function. The model estimates that the mass transfer coefficient is directly proportional to the square root of molecular diffusivity. That is

$$\textit{Mass transfer coefficient, } k_M = \sqrt{\frac{\textit{Molecular Diffusivity}}{\pi \, x \, \textit{Contact Time}}} \propto D^{\frac{1}{2}} \quad (4)$$

The Danckwert's or Random Surface Renewal.
Danckwerts P. V. (1951); Ind. Eng, Chem.; 43:1460

This model postulates that old and new surface elements have the same probability of renewal. It predicts that the surface age distribution function is $\Phi(t) = se^{-st}$ where s is the fractional rate of surface renewal and $0 \le t \le \theta_C$. The model estimates the mass transfer coefficient as being directly proportional to the square root of the product of molecular diffusivity and the fractional rate of surface renewal. That is

$$k_M = \sqrt{\textit{Molecular Diffusivity x Fractional Rate of Surface Renewal}}$$

$$= \sqrt{Ds} \propto D^{\frac{1}{2}} \tag{5}$$

The Perlmutter or Multiply Capacitance Model
(Perlmutter D. D. (1961); Chem. Eng. Sci; 16; 287)

This model postulates that the renewal of surface elements behaves like a system of n perfectly mixed tanks in series. That is, the surface residence time frequency distribution function, $f(t)$, is

$$f(t) = \frac{n^n t^{n-1}}{(n-1)!\theta^n} e^{-\frac{nt}{\theta}} \tag{6}$$

When n = ∞, equation (6) reduces to the Higbie model while, if n = 1, it reduces to the Danckwert's model. It estimates the mass transfer coefficient as

$$k_M = \sqrt{\frac{Molecular\,Diffusivity}{\pi\,x\,Contact\,Time}} \propto D^{\frac{1}{2}} \qquad see \quad (4)$$

The Surface Rejuvenation (Random Eddy) Model
(Harriott P. (1962); Chem. Eng. Sci; 17; 149)

This model states that surface elements are not completely renewed on exposure, as assumed for the Higbie, Danckwert's and Perlmutter models. Rather, they are rejuvenated as a result of dilution by new eddies. The model does not estimate the mass transfer coefficient. Rather it estimates the mass flux from a distribution of contact times.

The Surface Stretch Model
(Angelo J. B., E. N. Lightfoot and D. W. Howard (1966); AIChE J.; 12; 751)

This is an extension of the penetration theory and a surface renewal model in which the interfacial area, through which mass transfer occurs, changes, periodically, with time. Such situations arise when bubbles rise through denser liquid, or are formed at nozzles or when the surface of a liquid becomes rippled or wavy. This model predicts the mass transfer coefficient as (Treybal, 1980)

$$k_{L,avg} = \frac{\frac{A}{A_R}\sqrt{\frac{D_{AB}}{\pi\,\theta_R}}}{\sqrt{\int_0^{\theta/\theta_R}\left(\frac{A}{A_R}\right)^2 d\theta}} \propto D^{\frac{1}{2}} \tag{7}$$

27

where A = time dependent interfacial area, A_R = reference value of A for each situation, θ_R = a constant having the dimension of time, defined for each situation. For example, in drop formation, θ_R would be equal to the drop formation time.

Theories Incorporating Eddy Diffusion

Kichenevski - Kafarov Theory
(Kichenevski M. Kh. and Pamfilov A. P. (1949); Zh. Prikl. Khim.;
22; 1173;
Kafarov V. V.; Fundamentals of Mass Transfer, Vysshaya
Shkola, Moscow, 1962 (in Russian))

This model estimates the effective diffusivity, D_{EFF}, as the linear sum of the molecular, D_{AB}, and eddy, ε_D, diffusivities. That is

$$D_{EFF} = D_{AB} + \varepsilon_D \qquad (8)$$

Because of the intense mixing encountered in eddy diffusion, the model, also, assumes that the distribution of surface ages is narrow and has an average contact time of τ. Consequently, the mass transfer coefficient, k_M, is given by

$$k_M = 2 . \sqrt{\frac{D_{AB} + \varepsilon_D}{\pi . \tau}} \qquad (9)$$

When $\varepsilon_D > D_{AB}$,

$$k_M = 2 . \sqrt{\frac{\varepsilon_D}{\pi . \tau}} \qquad (10)$$

Turbulent Transfer Theory
(King C. J. (1966); Ind. Eng. Chem. Fundamentals; 5; 1)

This model, also, estimates the effective diffusivity, D_{EFF}, as the linear sum of the molecular diffusivity, D_{AB}, and eddy diffusivity, ε_D. That is

$$D_{EFF} = D_{AB} + \varepsilon_D \qquad see \qquad (8)$$

In addition, it estimates eddy diffusivity as a function of mixing or mass transfer distance, z. That is

$$\varepsilon_D = a z^n \qquad (11)$$

where *a* and *n* are constants. The theory reduces to, and predicts the mass transfer coefficient as,

- the penetration theory when *a* is very small or zero
- the Kichenevski - Kafarov theory when *n* = 0
- the film penetration theory when n = ∞

Example 2.11

Distinguish between molecular and eddy diffusion.

Answer

Molecular diffusion involves the net transfer of mass as a result of Brownian motion under a concentration difference driving force. No bulk or convective motion is involved.

Eddy diffusion involves the net transfer of mass as a result of turbulence and bulk or convective motion. Molecular diffusion still exists but is not significant in comparison to eddy diffusion.

Example 2.12

At 1 standard atmosphere, 100 C, the density of air is 0.9482 kg/m^3, its viscosity, 2.18 x 10^{-5} Ns/m^2, thermal conductivity, 0.0317 W/m^2 per K/m and specific heat at constant pressure, 1.047kJ/kg.K. Calculate, at 100 C, 1 atm,

1.　　the Prandtl number, Pr or N$_{Pr}$.
2.　　the thermal diffusivity, α
3.　　the kinematic or Stoke s viscosity, ν

If the Prandtl number is equal to the Schmidt number (Sc) at 100 C and the Schmidt number is independent of temperature, calculate the molecular diffusivity of air at 38 C if its viscosity at this temperature is 1.85 x 10^{-5} Ns/m^2 and its density is 1.14 kg/m^3.

Answer

You are expected to know, off head or, at least, recognise, these very common dimensionless groups. Hence for 1) the Prandtl number, Pr or N$_{Pr}$, is

$$Pr = \frac{Cp\,\mu}{k} = \frac{1.047 \times 1000 \times 2.18 \times 10^{-5}}{0.0317}, \frac{kJ}{kg.K} \frac{J}{kJ} \frac{kg}{m.s} \frac{m.K}{W} = 0.72$$

For 2) the thermal diffusivity, α

$$\alpha = \frac{k}{\rho Cp} = \frac{0.0317}{0.9482 \times 1047}, \frac{W}{m.K} \frac{m^3}{kg} \frac{kg.K}{J} = 3.193 \times 10^{-5}, \frac{m^2}{s}$$

29

For 3) the kinematic or Stoke's viscosity, ν

$$\alpha = \frac{\mu}{\rho} = \frac{2.18 \times 10^{-5}}{0.9482}, \frac{kg}{m.s} \frac{m^3}{kg} = 2.299 \times 10^{-5}, \frac{m^2}{s}$$

To calculate the molecular diffusivity at 38 C

Since $Sc = Pr = 0.72$ at 100 C and Sc is independent of temperature, at 38C, Sc will, also, be 0.72. Hence

$$Sc = \frac{\mu}{\rho D} = 0.72$$

or $\qquad D = \frac{\mu}{0.72\rho} = \frac{1.85 \times 10^{-5}}{0.72 \times 1.14}, \frac{kg}{m.s} \frac{m^3}{kg} = 2.254 \times 10^{-5} \frac{m^2}{s}$ Ans

Example 2.13

Pure SO_2, in the gas phase, is in contact, at atmospheric pressure and 20 C. with a pure quiescent liquid layer of considerable depth into which it is being absorbed. After one hour, the concentration of SO_2 in a sample, withdrawn from a point 5mm below the water surface, was 1.04 kmol/m³. Calculate the molecular diffusivity of SO_2 assuming that the penetration theory applies. You are given that

$$\frac{C_A - C_{A0}}{C_{Ai} - C_{A0}} = 1 - erf\left(\frac{z}{\sqrt{4 D_{AB} t}}\right) \qquad (a)$$

where $\qquad erf\ x = \frac{2}{\sqrt{\pi}} \int_0^x e^{u^2} du \qquad (b)$

and

x	0.1	0.2	0.3	0.4	0.5
$erf\ x$	0.113	0.223	0.329	0.429	0.521

The solubility of SO_2 in water at 20 C is 1.54 kmol/m³.

Answer

The physical situation may be represented as follows:

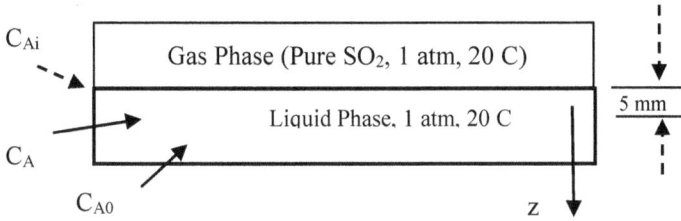

C_{Ai} is the concentration of SO_2 at the gas-liquid interface. It is, also, equal to the solubility of SO_2 (since this is the concentration that is in equilibrium at 20 C with 1 atm of pure SO_2 gas) = 1.54 kmol/m^3.

Since the bulk of the liquid layer is pure liquid, without any SO_2, C_{A0} = 0. At 5 mm below the interface, C_A = 1.04 kmol/m^3, z = 0.005 m, t = 3600 s. Substituting these values into equation (a), we get

$$\frac{1.04}{1.54} = 1 - erf\left(\frac{0.005}{\sqrt{4 x D_{AB} x 3600}}\right) = 1 - erf\left(\frac{0.005}{120\sqrt{D_{AB}}}\right)$$

That is

$$erf\left(\frac{0.005}{120\sqrt{D_{AB}}}\right) = 1 - 0.675 = 0.325 = erf\ x \qquad (1)$$

where

$$x = \frac{0.005}{120\sqrt{D_{AB}}} \qquad (2)$$

To obtain x we need to plot $erf\ x$ versus x and to read off the required x from the graph. This is done in the figure below. From it, $x = 0.294$. Thus

$$\sqrt{D_{AB}} = \frac{0.005}{120 x 0.294} = 1.4172 x 10^{-4}\ or\ D_{AB} = 2.009 x 10^{-8}, \frac{m^2}{s}\ Ans.$$

Plot of Erf (x) Versus x

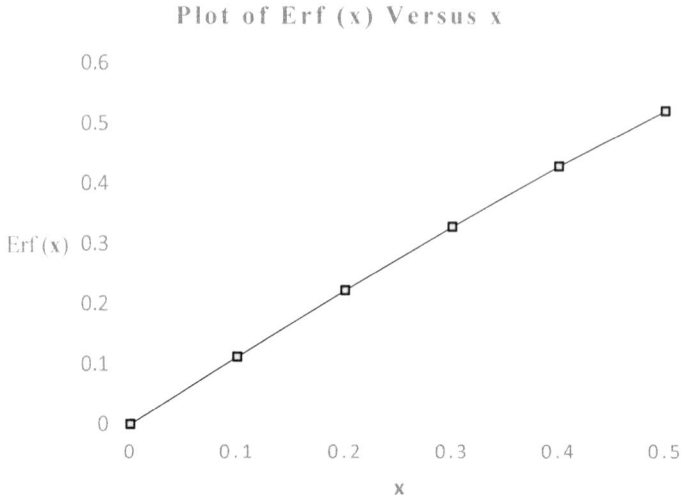

References for Chapter Two

1.	Angelo J. B., E. N. Lightfoot and D. W. Howard (1966); AIChE J;**12**; 751

2.	Danckwerts P. V.(1951); Lid. Eng. Chem.; **43**; 1460

3.	Dean Lange's Handbook of Chemistry, McGraw-Hill Book Co., N.Y. 1973

4.	Fuller E. N., Schettler P. D., and Giddings J. D. (1966); *A New Method for the Prediction of Gas Phase Diffusion Coefficients*; Ind. Eng. Chem., **58**; May 19.

5.	Harriott P. (1962); Chem. Eng. Sci; **17**; 149

6.	Higbie R. (1935): Trans. Am. Inst. Chem. Engrs; **31**, 365

7.	Hirschfelder, Bird and Spotz in Colburn A. P. (1933); Trans. AIChE; **29**; 174

8.	Kafarov V. V.; Fundamentals of Mass Transfer, Vysshaya Shkola, Moscow; 1965.

9.	King C. J. (1966); Ind. Eng. Chem., Fundamentals; **51**; 1; Am. Chem. Soc., Wash. D.C., USA.

10.	Kishenevski M, Pamfilov A. P. (1949); Zh. Prikl. Khim; **22**;1173

11.	Perlmutter D. D. (1961); Chem. Eng. Sci.; **16**; 287

12.	Perry R. and Green D; Chemical Engineers' Handbook, 6th. Edn.; McGraw-Hill Book Co., N.Y., USA, 1984.

13.	Sawistowski H; Class Notes in Mass Transfer; Imp. Coll., London; 1965.

14.	Toor H. L. and Marchello J. M. (1958); AIChE,4; 97

15.	Treybal R. E.; Mass Transfer Operations; 3rd. Edn.; McGraw-Hill Book Co., N.Y., USA, 1980.

16.   Whitman W. G. (1923); Chem. Met. Eng.; **29**; 147

17.   Wilke and Chang (1923); J. AIChE, **1**; 264.

CHAPTER THREE:
THE MASS TRANSFER COEFFICIENT

The concept of the mass transfer coefficient is useful in analysing mass transfer operations. This is, mainly, because it is a performance parameter since if

$$Average\ molar\ flux = constant \times concentraion\ driving\ force \quad (3.1)$$

$$then \quad constant = \frac{Average\ molar\ flux}{Concentration\ driving\ force} \quad (3.2)$$

This *constant* is defined as the mass transfer coefficient.

Thus, the mass transfer coefficient, in any system, is a measure of how much of the average molar flux can be obtained for each unit of concentration driving force. Unlike the molecular diffusivity, the mass transfer coefficient is associated with bulk motion of the phases. This bulk motion implies the existence of boundary layers in each phase next to the interface between the phases. Each phase boundary layer, therefore, will present a different level of resistance to mass transfer and will, therefore, have its own mass transfer coefficient.

Thus, in dealing with mass transfer problems in which the mass transfer coefficient is to be used, it is necessary to know

i)   the mass transfer coefficient in each phase
ii)   the overall mass transfer coefficient of the system

The ten problems in this Chapter seek to illustrate the definitions, and the estimation, of both the individual phase, and overall, mass transfer coefficients for gas and liquid phases, using established methods. The phases to be examined include those in laminar and turbulent flow as well as those that are continuous or disperse,

Example 3.1

a).   Define the concept of the mass transfer coefficient.
b).   Is it related to the diffusion coefficient?
c).   10% SO_2, by volume in air, in contact with water at atmospheric

pressure and at 50 C, contains 0.4% SO_2 (density = 990 kg/m³). The equilibrium partial pressure of SO_2 at the air-water interface is 70.68 mm. Hg. The air based film mass transfer coefficient is 1.566 x 10^{-9} kmol/m².s per N/m². Determine the molar flux of SO_2 across the air-water interface. Assume that only SO_2 diffuses across the interface. Take 1 atm. pressure = 76 cm Hg = 1.013 x 10^5 N/m².

Answer

Part a

The mass transfer coefficient is defined as the ratio of the molar flux to the concentration driving force for mass transfer. That is

$$Mass\ Transfer\ Coefficient\ =\ \frac{Molar\ flux}{Concentration\ driving\ force} \tag{1}$$

The mass transfer coefficient is, always, defined with respect to a particular phase of interest. Thus, in the gas phase, the mass transfer coefficient is designated as either

$$k_G\ =\ \frac{Molar\ flux}{Partial\ pressure\ driving\ force} \tag{2}$$

with units of moles per m² per second per N/m². The gas phase mass transfer coefficient can, also, be expressed in mole fractions as

$$k_y\ =\ \frac{Molar\ flux}{Mole\ fraction\ driving\ force} \tag{3}$$

with units of moles per m² per second per unit mole fraction. In terms of concentration, the gas phase mass transfer coefficient can, also, be expressed as

$$k_C\ =\ \frac{Molar\ flux}{Concentration\ driving\ force} \tag{4}$$

with units of moles per m² per second per unit concentration difference.

The relationships between these various definitions of the mass transfer coefficient in the gas phase can be derived as follows.

For any gas, the mole fraction, y_A = P_A / P_T where P_A is the partial pressure of component A, and P_T is the total pressure of the system. For an ideal gas, $P_A.V = n_A.RT$ and since concentration is in moles per unit volume,

$$C_A = \frac{n_A}{V} = \frac{P_A}{RT} \tag{5}$$

By defining molar flux in terms of partial pressure, mole fraction and molar concentration, respectively, and using the appropriate mass transfer coefficient, we get

$$N_A = k_G \left(P_{A1} - P_{A2}\right) = k_y \left(y_{A1} - y_{A2}\right) = k_C \left(C_{A1} - C_{A2}\right) \tag{6}$$

When we express y_A and C_A in terms of P_A, the molar flux equation becomes

$$N_A = k_G \left(P_{A1} - P_{A2}\right) = \frac{k_y}{P_T} \left(P_{A1} - P_{A2}\right) = \frac{k_C}{RT} \left(P_{A1} - P_{A2}\right) \tag{7}$$

It is now easy to see that

$$k_G = \frac{k_y}{P_T} = \frac{k_C}{RT} \tag{8}$$

Note that a k_P or a k_g based on *mm. Hg* or atmospheres driving force as well as a k_Y based on mole ratio, Y_A, driving force are, also, used.

In the liquid phase, the mass transfer coefficient is designated, as in the gas phase, either as

$$k_y = \frac{Molar\ flux}{Mole\ fraction\ driving\ force} \qquad as\ in\ (3)$$

with units of moles per m^2 per second per unit mole fraction or as

$$k_C = \frac{Molar\ flux}{Concentration\ driving\ force} \qquad as\ in\ (4)$$

with units of moles per m^2 per second per unit concentration difference. They are related as

$$N_A = k_x \left(x_{A1} - x_{A2}\right) = k_L \left(C_{A1} - C_{A2}\right) = \frac{k_x}{C_T} \left(C_{A1} - C_{A2}\right) \tag{9}$$

from which it can be seen that

$$k_L = \frac{k_x}{C_T} \tag{10}$$

Part b

The relationship between the mass transfer coefficient, k, and the molecular diffusivity, D, may be illustrated as shown below.

Generally, the effective diffusivity is, usually, some summation of the molecular and eddy diffusivities. The simplest relationship we can use at this time is that the effective diffusivity is proportional to some linear sum of the molecular and eddy diffusivities. That is

$$D_{eff} \propto (D + \varepsilon_D) \tag{11}$$

where D_{eff} = effective diffusivity, ε_D = eddy diffusivity and D = molecular diffusivity.

For highly agitated interfaces, the eddy diffusivity is much larger than the molecular diffusivity, i.e. $\varepsilon_D >>> D$ so that D can be ignored with respect to ε_D and $D_{eff} = \varepsilon_D$. Since the mass transfer coefficient, k, is proportional to the effective diffusivity,

$$k \propto D_{eff}^n = \varepsilon_D^n \tag{12}$$

where the exponent, n, is, as yet, unknown. This means, however, that in such cases, k is independent of D.

For quiescent interfaces for which $D >>> \varepsilon_D$, ε_D can be ignored with respect to D so that $D_{eff} = D$. Thus

$$k \propto D_{eff}^n = D^n \tag{13}$$

and n is found, experimentally, to be equal to 2/3 while various theories of mass transfer predict values of 0.5 or 1.0.

When $D = \varepsilon_D$, ε_D is predicted to be proportional to z^m. Since $D_{eff} \propto (D + \varepsilon_D)$, and $k \propto D_{eff}^n$, then

$$k \propto (D^n + a z^m) \tag{14}$$

where n, a and m are constants.

Equations (12), (13) and (14) thus show that k and D are related and in such a way that, in highly agitated interfaces, k is independent of D, in quiescent interfaces, k is proportional to D^n while for interfaces in which eddy and molecular diffusivity are of, more or less, equal magnitude, $k \propto (D^n + a z^m)$, where z is the diffusion distance from the interface into the medium.

Part c

Since the air and water interface is at the same temperature and pressure, the transfer is isothermal and we can estimate the molar flux using standard equations. Thus

Molar flux, from equation (6)

$$N_A = k_G \left(P_A - P_{Ai} \right) \qquad\qquad from \quad (6)$$

where

P_A = partial pressure of SO_2 in the gas phase
= 0.1 x (760/760) x 1.013 x 10^5
= 10130 N/m²

P_{Ai} = partial pressure of SO_2 in the gas phase, at the interface
= 70.68 mm Hg = (70.68/760) x 1.013 x 10^5
= 9420.9 N/m²

Substituting these and the given value of k_G = 1.566 x 10^{-9}, kmol/m².s/ N/m² into the molar flux equation, we get that

$$N_A = 1.566 \times 10^{-9} \left(10130 - 9420.9 \right), \frac{kmol}{m^2.s} \frac{m^2}{N} \frac{N}{m^2}$$

$$= 1.111 \times 10^{-6} \frac{kmol}{m^2.s} \quad Ans.$$

Example 3.2

The molar flux of ammonia through a film of stagnant gas, 0.5 mm thick, when the concentration change across the film is 10 to 5 % ammonia by volume, is 2.05 x 10^{-4} kmol/m².s. The total pressure is 207 kN/m² and the temperature is 54 C. Calculate the mass transfer coefficient based on a) partial pressure driving force, b) mole fraction driving force and c) concentration driving force.

Answer

The mass transfer coefficient, in the gas phase, can be defined as

$$k_G = \frac{N_A}{\Delta P_A} \quad or \ as \quad k_y = \frac{N_A}{\Delta y_A} \quad or \ as \quad k_C = \frac{N_A}{\Delta C_A} \qquad (1)$$

where ΔP_A, Δy_A and ΔC_A are, respectively, the partial pressure, mole fraction and concentration driving force for mass transfer and N_A is the molar flux of component A.

The driving force change of 10 to 5% by volume is equivalent to

1. A partial pressure driving force change of

$$\Delta P_A = \left(0.1 - 0.05 \right) \times 207 = 10.35 \frac{kN}{m^2} \qquad (2)$$

2. A mole fraction driving force change of

$$\Delta y_A = \left(0.1 - 0.05 \right) = 0.05 \qquad (3)$$

and

3.   A concentration driving force change of

$$\Delta C_A = \frac{\Delta P_A}{RT} = \frac{10.35}{8.314 x (273 + 54)} \frac{kN}{m^2} \frac{kmol.K}{kN.m} \frac{1}{K}$$

$$= 3.807 x 10^{-3} \frac{kmol}{m^3} \tag{4}$$

Hence, since $N_A = 2.05 \times 10^{-4}$ kmol/m².s, given

$$k_G = \frac{N_A}{\Delta P_A} = \frac{2.05 x 10^{-4}}{10.35}, \frac{kmol}{m^2.s} \frac{m^2}{kN}$$

$$= 1.981 x 10^{-5} \frac{kmol}{m^2.s} per \frac{kN}{m^2} \qquad Ans$$

$$k_y = \frac{N_A}{\Delta y_A} = \frac{2.05 x 10^{-4}}{0.05}, \frac{kmol}{m^2.s} \frac{1}{mole\ fraction}$$

$$= 4.100 x 10^{-3} \frac{kmol}{m^2.s} per\ unit\ mole\ fraction \qquad Ans$$

$$k_C = \frac{N_A}{\Delta C_A} = \frac{2.05 x 10^{-4}}{3.807 x 10^{-3}}, \frac{kmol}{m^2.s} \cdot \frac{m^3}{kmol}$$

$$= 0.054 \frac{kmol}{m^2.s} per \frac{kmole}{m^3} = 0.054 \frac{m}{s} \qquad Ans$$

Example 3.3

Water flows at 30 C through a 50 mm ID pipe at an average velocity of 3.05 m/s. If a 61 cm section of the pipe is replaced with a 50 mm ID tube of solid sodium chloride of the same length, determine the mass transfer coefficient of sodium chloride in water using

(a)   the Reynolds's analogy

$$\frac{k_L}{u} = \frac{R}{\rho u^2} = Re^{-0.25} \tag{1}$$

(b)   the Chilton - Colburn analogy

$$j_D = \frac{k_L}{u} Sc^{2/3} = 0.027 Re^{-0.2} \tag{2}$$

where R is the shear stress; Re, the Reynold's number = $\rho du/\mu$; Sc, the

Schmidt number = $\mu/\rho D$. Take the density of water, $\rho = 1000$ kg/m^3; its viscosity, $\mu = 0.860$ x 10^{-3}, Pa.s and the diffusivity of sodium chloride in water, $D = 1.26$ x 10^{-3}, m^2/s. [Na = 23, Cl = 35.5, H = 1, O = 16].

Answer

Since $u = 3.05$ m/s and $d = 0.05$ m

$$\text{Re} = \frac{\rho d u}{\mu} = \frac{1000 x 0.05 x 3.05}{0.860 x 10^{-3}}, \frac{kg}{m^3} \cdot \frac{m}{1} \cdot \frac{m}{s} \cdot \frac{m.s}{kg} = 177{,}325.6 \quad (3)$$

$$Sc = \frac{\mu}{\rho D} = \frac{0.860 x 10^{-3}}{1000 x 1.26 x 10^{-9}}, \frac{kg}{m.s} \cdot \frac{m^3}{kg} \cdot \frac{s}{m^2} = 682.5 \quad (4)$$

a)   By Reynolds's analogy

$$k_L = u.\text{Re}^{-0.25} = 3.05 x \left(177{,}325.6\right)^{-0.25}, \frac{m}{s} = 0.149\frac{m}{s} \quad Ans$$

(b)   By the Chilton - Colburn analogy

$$k_L = \frac{u}{Sc^{2/3}} x \frac{0.027}{\text{Re}^{0.2}} = \frac{3.05}{\left(682.5\right)^{2/3}} x \frac{0.027}{\left(177{,}325.6\right)^{0.2}}, \frac{m}{s}$$

$$= 0.03935 x 0.002408 = 9.475 x 10^{-5}, \frac{m}{s} \quad Ans$$

Note the great discrepancy in the values of k_L obtained by the two analogies for this problem. The Chilton-Colburn analogy, however, is found to give more accurate predictions.

Example 3.4

100 g/min of air is passed up a wetted wall column of diameter, 2.67 cm, down which n-butyl alcohol (n-BA) is flowing. The system temperature is 30 C. It is desired to estimate the air based, film, mass transfer coefficient using the Chilton-Colburn analogy, given, for diffusion through a stagnant medium, by

$$j_D = \frac{k_L}{u} \cdot \frac{P_{BM}}{P_T} . Sc^{2/3} = 0.027 \text{Re}^{-0.2} \quad (1)$$

where the Reynolds's number, Re = $\rho u d/\mu$; the Schmidt number, Sc = $\mu/\rho D$; k_L = the mass transfer coefficient, m/s; u = the mean velocity of the air-n-BA stream, m/s; P_T = total system pressure; P_{BM} = the log mean partial pressure of the non-diffusing component; μ = the mean viscosity of the air-n-BA stream; ρ = the mean density of the air-n-BA stream; D

= diffusivity of n-BA in air and d = diameter of the column.

If P_T = 820 mm. Hg; P_{BM} = 799 mm. Hg; μ = 1.97 x 10^{-4}, g/cm.s; ρ = 1.18 x 10^{-3}, g/cc; D = 0.0803 cm²/s, determine k_L and k_G for the system, given that the mean mass flux of the air-n-BA stream is 0.305 g/cm².s. Note that 760 mm Hg = 1.013 x 105 N/m² and that R = 82.06 atm.cm³/gmol.K.

Answer

$$Sc = \frac{\mu}{\rho D} = \frac{1.97 x 10^{-4}}{1.18 x 10^{-3} x 0.0803}, \frac{g}{cm.s} . \frac{cm^3}{g} . \frac{s}{cm^2} = 2.079 \qquad (2)$$

Since mean mass flux = $\rho\, u$ = 0.305 g/cm².s

$$u = \frac{3.05}{\rho} = \frac{3.05}{1.18 x 10^{-3}}, \frac{g}{cm^2.s} . \frac{cm^3}{g} = 258.5, \frac{cm}{s} \qquad (3)$$

$$Re = \frac{\rho\, d\, u}{\mu} = \frac{1.18 x 10^{-3} x 2.67 x 258.5}{1.97 x 10^{-4}}, \frac{g}{m^3} . \frac{cm}{1} . \frac{cm}{s} . \frac{cm.s}{g} = 4134.2 \qquad (4)$$

Hence, from the Chilton-Colburn equation

$$k_L = \frac{u}{Sc^{2/3}} . \frac{P_T}{P_{BM}} . \frac{0.027}{Re^{0.2}} = \frac{258.5}{(2.079)^{2/3}} x \frac{820}{799} x \frac{0.027}{(4134.2)^{0.2}}, \frac{m}{s}$$

$$= 158.693 x 1.026 x 0.005106 = 0.831, \frac{m}{s} \qquad Ans$$

Since k_L = $k_G RT$ and T = 273 + 30 = 303 K, then

$$k_G = \frac{k_L}{RT} = \frac{0.831}{82.06 x 303}, \frac{cm}{s} . \frac{gmol.K}{atm.cm^3} . \frac{1}{K} = 3.34 x 10^{-5}, \frac{gmol}{cm^2.s.atm} \qquad Ans$$

Example 3.5

Explain why it is necessary to define and use overall mass transfer coefficients. Hence show that in any one phase, the overall mass transfer coefficient, based on that phase, is given by

$$\frac{1}{K_G} = \frac{1}{k_G} + \frac{m_1}{k_L} \quad and \quad \frac{1}{K_L} = \frac{1}{k_L} + \frac{1}{m_2 k_G} \qquad (1)$$

where K_G and K_L are the overall mass transfer coefficients, based on the gas and liquid phases, respectively, and k_G and k_L the individual film coefficients in those phases. Explain m_1 and m_2.

Answer

The molar flux is defined in each phase as $k_G.\Delta P_G = k_L.\Delta C$ where ΔP_G and ΔC are the pressure and concentration driving forces, respectively. The molar flux calculated for each phase does not differ from that calculated for the other and each represents, accurately, the molar flux for the system. To use a mass transfer coefficient for one phase, which does not incorporate both phase transfer coefficients, would, however, be unrealistic, in commercial equipment. This is because one phase transfer coefficient may be insensitive to factors which influence the other. Using it alone would lead to results, especially in sizing the equipment, which do not represent the real or total situation.

To show that

$$\frac{1}{K_G} = \frac{1}{k_G} + \frac{m_1}{k_L} \quad and \quad \frac{1}{K_L} = \frac{1}{k_L} + \frac{1}{m_2\,k_G}$$

consider the figures below which illustrate the equilibrium curves for the mass transfer of a component, A, between a light and a heavy phase across an interface. P_A represents the partial pressure of component, A, in the lighter phase. The equilibrium curves can be, generally, of three types. In Type 1, component A is highly soluble in the heavy phase. In Type 2, component A has low solubility in the heavy phase while in Type 3, the solubility of A is comparable in both phases.

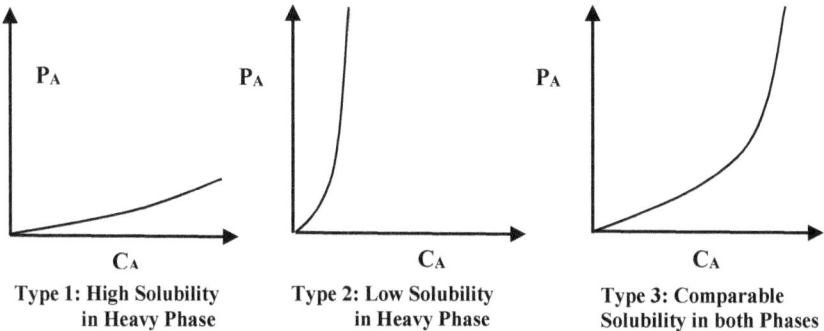

| Type 1: High Solubility in Heavy Phase | Type 2: Low Solubility in Heavy Phase | Type 3: Comparable Solubility in both Phases |

Using the Type 3 curve as the most general, consider any operating composition in a mass transfer system such as D in the figure below.

Such a composition would have a partial pressure P_{AD} in the gas or light phase corresponding to C_{AD} in the liquid or heavy phase. If the film mass transfer coefficients in the liquid and gas phases are given by k_L and k_G, respectively, the equilibrium condition that can be reached from the composition at D would be given at E by the line from D, of slope, -k_L/k_G, to the equilibrium curve. This is because the molar flux, N_A, is given by $N_A = k_G (P_{AD} - P_{Ai})$ in the gas phase and by $N_A = k_L(C_{Ai} - C_{AD})$ in the liquid phase.

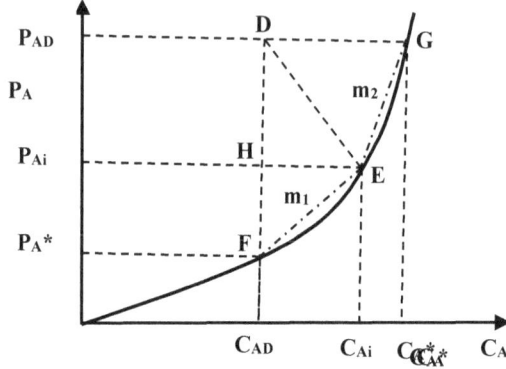

Since
$$k_G \left(P_{AD} - P_{Ai}\right) = N_A = k_L \left(C_{Ai} - C_{AD}\right) \tag{2}$$

$$\frac{P_{AD} - P_{Ai}}{C_{AD} - C_{Ai}} = -\frac{k_L}{k_G} = \frac{DH}{HE} \tag{3}$$

The equilibrium composition at E will be (P_{Ai}, C_{Ai}). If a mass transfer coefficient is defined which takes into account the transfer in both phases but is calculated on the basis of the driving force in one phase, then $N_A = K_G \left(P_{AD} - P_A^*\right)$ in the gas phase and $N_A = K_L \left(C_A^* - C_{AD}\right)$ in the liquid phase. And since $K_G \left(P_{AD} - P_A^*\right) = N_A = K_L \left(C_A^* - C_{AD}\right)$

$$\frac{P_{AD} - P_A^*}{C_{AD} - C_A^*} = -\frac{K_L}{K_G} = \frac{DF}{DG} \tag{4}$$

Note that $-K_L/K_G = DF/DG$ represents the average slope of the equilibrium curve within the concentration range of interest. The derivation of the relationships between K_L and K_G and between k_L and k_G is mere algebra and is illustrated below.

$$P_{AD} - P_A^* = (P_{AD} - P_{Ai}) + (P_{Ai} - P_A^*) = \frac{N_A}{K_G} \tag{5}$$

$$C_A^* - C_{AD} = (C_A^* - C_{Ai}) + (C_{Ai} - C_{AD}) = \frac{N_A}{K_L} \tag{6}$$

But

$$P_{AD} - P_{Ai} = \frac{N_A}{k_G}; \qquad C_{Ai} - C_{AD} = \frac{N_A}{k_L} \qquad (7)$$

And

$$P_{Ai} - P_A^* = m_1 \left(C_{Ai} - C_{AD} \right); \qquad P_{AD} - P_{Ai} = m_2 \left(C_A^* - C_{Ai} \right) \qquad (8)$$

Substituting from (7) and (8) in (5) for partial pressures

$$\frac{N_A}{k_G} + m_1 \frac{N_A}{k_L} = \frac{N_A}{K_G} \quad or \quad \frac{1}{K_G} = \frac{1}{k_G} + \frac{m_1}{k_L} \qquad (9)$$

Similarly, from (7) and (8) in (6) for concentration

$$\frac{N_A}{m_2 k_G} + \frac{N_A}{k_L} = \frac{N_A}{K_L} \quad or \quad \frac{1}{K_L} = \frac{1}{k_L} + \frac{1}{m_2 k_G} \qquad (10)$$

Note that m_1 and m_2 are the slopes of the lines which join the real equilibrium composition (C_{Ai} and P_{Ai}) to the hypothetical equilibrium compositions (C_A^* and P_A^*) on the basis of which the overall mass transfer coefficients are defined.

Example 3.6

The overall mass transfer coefficient, based on the gas phase, in a tower for the absorption of ammonia into water from an air stream, was found to be 2.56 x 10^{-9} kmol/m².s per N/m². The tower was operated at 2 atm. and 15.6 C.

At the point of interest, the air stream contained 1% by volume of ammonia and the water was pure. If the gas phase offered 85% of the resistance to mass transfer, calculate

a) the gas film coefficient, k_G
b) the liquid film coefficient, k_L
c) the interfacial equilibrium concentrations, P_{Ai} and C_{Ai}

You are given that for dilute solutions of ammonia in water at 15.6 C, the equilibrium partial pressure of ammonia is given by $P_{Ai} = 0.0156 C_{Ai}$, where P_{Ai} is in atmospheres and C_{Ai} is the concentration of ammonia in water, kmol/m³.

Answer

At the point of interest

$$Total\ Resis\tan ce\ to\ Mass\ Transfer = Resis\tan ce\ in\ the\ gas\ phase$$
$$+ Resis\tan ce\ in\ the\ liquid\ phase \qquad (1)$$

That is

$$\frac{1}{K_G} = \frac{1}{k_G} + \frac{m}{k_L} \qquad (2)$$

a). To evaluate the gas film coefficient

Since $K_G = 2.56\ x\ 10^{-3}$ kmol/m².s per N/m² and the gas phase resistance = 85% of total resistance, then

$$\frac{1}{k_G} = 0.85 x \frac{1}{K_G} \qquad (3)$$

That is

$$k_G = \frac{K_G}{0.85} = \frac{2.56x10^{-9}}{0.85}, \frac{kmol}{m^2.s}.\frac{m^2}{N} = 3.012\,x10^{-9}, \frac{kmol}{m^2.s}.\frac{m^2}{N}$$

$$= 3.012\,x10^{-9}, \frac{kmol}{m^2.s}.\frac{m^2}{N}\,x\,1.013x10^5, \frac{N}{m^2}.\frac{1}{atm}$$

$$= 3.051x10^{-4}, \frac{kmol}{m^2.s.atm} \quad Ans$$

b) To evaluate the liquid film coefficient

Since the gas phase resistance is 85% of total resistance, then 0.15% of total resistance must be in the liquid phase, assuming no surface resistance. That is

$$\frac{m}{k_L} = 0.15 x \frac{1}{K_G} \qquad (4)$$

But $m = 0.0156\ atm.\ per\ \dfrac{kmol}{m^3}$

$$= 0.0156\frac{atm.m^3}{kmol}\,x\,1.013x10^5, \frac{N}{m^2}.\frac{1}{atm} = 1580.28\frac{N}{m^2}\,per\,\frac{kmol}{m^3} \qquad (5)$$

Thus

$$k_L = \frac{m\,K_G}{0.15} = \frac{1580.28 \times 2.56 \times 10^{-9}}{0.15}, \frac{N}{m^2} \cdot \frac{m^3}{kmol} \cdot \frac{kmol}{m^2.s} \cdot \frac{m^2}{N}$$

$$= 2.697 \times 10^{-5}, \frac{kmol}{m^2.s} \; per \; \frac{N}{m^2} \qquad Ans$$

c) To determine the interfacial equilibrium concentrations

Since the equilibrium line is given by $P_{Ai} = mC_{Ai}$ a line from the point (P_A, C_A) will intersect the equilibrium line at (P_{Ai}, C_{Ai}) with a slope of $-k_L/k_G$. An analytical or graphical solution will give the desired result. Graphiocal solutions are, however, tedious and rarely used these days especially where analytical solutions are available.

Analytically, the equation of the line from (P_A, C_A) to (P_{Ai}, C_{Ai}), of slope $-k_L/k_G$, will be

$$\frac{P_A - P_{Ai}}{C_A - C_{Ai}} = -\frac{k_L}{k_G} = -\frac{2.697 \times 10^{-5}}{3.051 \times 10^{-4}}, \frac{m}{s} \cdot \frac{m^2.s}{kmol} \cdot \frac{atm}{1} = -0.088 \frac{atm.m^3}{kmol} \quad (6)$$

Since $P_A = 0.01 \times 2 = 0.02$ atmospheres and $C_A = 0$ then

$$\frac{P_A - P_{Ai}}{C_A - C_{Ai}} = -0.088 \frac{atm.m^3}{kmol} = \frac{0.02 - P_{Ai}}{0 - C_{Ai}}$$

or

$$P_{Ai} = 0.02 - 0.088 C_{Ai} \qquad (7)$$

But $P_{Ai} = 0.0156 C_{Ai}$. Therefore $P_{Ai} = 0.0156 C_{Ai} = 0.02 - 0.088 C_{Ai}$ from which

$$C_{Ai} = \frac{0.02}{0.1036} = 0.193, \frac{kmol}{m^3} \qquad Ans$$

and

$$P_{Ai} = 0.0156 \times 0.193 = 3.01 \times 10^{-3} \; atm \quad Ans$$

Graphically, equation (7) and the equilibrium curve can be plotted on a graph and their point of intersection will give P_{Ai}, C_{Ai}. By choosing different values of C_A, P_A is calculated for the two equations as shown in the table below.

C_{Ai}	$P_{Ai} = 0.02 - 0.088C_{Ai}$	$P_{Ai} = 0.0156C_{Ai}$
0	0.0200	0
0.05	0.0156	0.00078
0.10	0.0112	0.00156
0.15	0.0068	0.00234
0.20	0.0024	0.00312
0.25	0.0020	0.00390

These are plotted as shown below and the answer will be seen, from the intersection of the two lines, to be $P_{Ai} = 0.003$ atm and $C_{Ai} = 0.193$ kmol/m^3.

Fig. 3.6: Graphical Solution of Example 3.6c

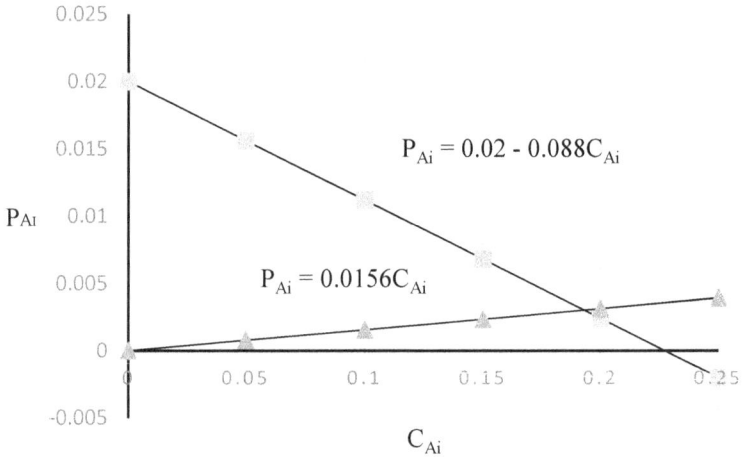

Example 3.7

10% SO_2, by volume, in air, is in contact with water containing 0.4% SO_2 (solution density = 990 kg/m^3) at 50 C and 1 atm. pressure. The overall mass transfer coefficient, based on gas concentrations, was $K_G = 7.36 \times 10^{-10}$ kmol/m^2.s per N/m^2 of which 47% of the total diffusional resistance lay in the gas phase. The equilibrium data at 50 C are given as

kg SO_2 per 100 kg water	0.2	0.3	0.5	0.7
Partial pressure SO_2, mm Hg	29	46	83	119

48

Calculate
- a) the individual film coefficients, k_G and k_L
- b) the interfacial composition in both phases and
- c) the overall mass transfer coefficient based on liquid concentrations (mol/volume)

Answer

a) Estimating individual film coefficients

The gas phase based overall mass transfer coefficient is related to the film coefficients as follows

$$\frac{1}{K_G} = \frac{1}{k_G} + \frac{m_1}{k_L} \tag{1}$$

That is: Total resistance = resistance in gas + resistance in liquid
Since gas phase resistance is 47% of total resistance, 53% of total resistance must be in the liquid phase.

Hence

$$\frac{1}{k_G} = 0.47 x \frac{1}{K_G} \tag{2}$$

That is

$$k_G = \frac{K_G}{0.47} = \frac{7.36x10^{-10}}{0.47} , \frac{kmol}{m^2.s} \frac{m^2}{N} = 1.566 x 10^{-9}, \frac{kmol}{m^2.s} \frac{m^2}{N} \quad Ans$$

For the liquid phase

$$\frac{m}{k_L} = 0.53 x \frac{1}{K_G} \quad or \quad k_L = \frac{m K_G}{0.53} \tag{3}$$

To evaluate k_L, m has to be determined. To determine m, the slope of the equilibrium line, we need to plot the given equilibrium data. The units with which the data are plotted will depend on the units of k_G and k_L. For example, if we use the expression, $N_A = k_G \Delta P_A$, k_G will have the units of $kmol/m^2.s$ per N/m^2. Similarly, if we use $N_A = k_L \Delta C_A$, ΔC_A will have the units of $kmol/m^3$ and k_L the units of m/s.

The given data are, therefore, recalculated, as below, in conformity with the requirements above. To convert partial pressure, mm Hg to N/m^2, remember that 1 atm. = 1.013×10^5 N/m^2 = 760 mm Hg. Hence, at any pressure, P, mm Hg, the pressure in Pascals, P_A, N/m^2, is given by

49

$$P_A = \frac{P}{760} x\ 1.013 x 10^5, \frac{N}{m} \tag{4}$$

To convert from the mass ratio given, kg SO_2 /100 kg H_2O, to molar concentration, kmol/m^3, we shall, first, convert the given mass ratio, to mass concentration, kg SO_2/kg solution. Thereafter, we shall convert this mass concentration to molar concentration as follows. Let X = mass ratio SO_2 to H_2O. Then

$$X = \frac{kg\ SO_2}{kg\ H_2O} = \frac{1}{100} x \frac{kg\ SO_2}{100\,kg\ H_2O} \tag{5}$$

The mass fraction, x, is given by

$$x = \frac{X}{1+X} = \frac{1}{100} x \frac{kg\ SO_2/kg\ H_2O}{1 + kg\ SO_2/kg\ H_2O} = \frac{kg\ SO_2}{kg\ SO_2 + kg\ H_2O}$$

$$= \frac{kg\ SO_2}{kg\ Solution} \tag{6}$$

Mass concentration of SO_2 in H_2O = $\dfrac{kg\ SO_2}{m^3\ of\ Solution}$

$$= \frac{kg\ SO_2}{kg\ Solution} x \frac{kg\ of\ solution}{m^3\ of\ Solution} = \frac{kg\ SO_2}{kg\ Solution} x\,density\ of\ solution \tag{7}$$

Molar concentration of SO_2 in H_2O = $\dfrac{Mass\ concentraion}{Molecular\ weight}$

$$= \frac{kg\ SO_2}{kg\ Solution} x \frac{Density\ of\ solution}{Molecular\ weight\ of\ SO_2}$$

$$= \frac{X}{1+X} . \frac{Density\ of\ solution}{Molecular\ weight\ of\ SO_2} = \frac{x\rho}{64}, \frac{kmol}{m^3} \tag{8}$$

Since the solution is very dilute, ρ can be regarded as constant and equal to 990 kg/m^3. Taking SO_2 to be component A and applying these conversion formulae to the given equilibrium data, we get

P_A, N/m^2	1333	6131	11063	15862
C_A, kmol/m^3	0.015	0.046	0.077	0.108

These are plotted in the Figure below from which the slope of the curve, m, is obtained as 156225.8 N/m² per kmol/m³. We can now evaluate k_L as

$$k_L = \frac{mK_G}{0.53} = \frac{156225.8 \; x \; 7.36^{-10}}{0.53} = 2.1695 \; x \; 10^{-4}, \frac{m}{s}$$

Plot Of The Equilibrium Line (Assuming Very Dilute Solution)

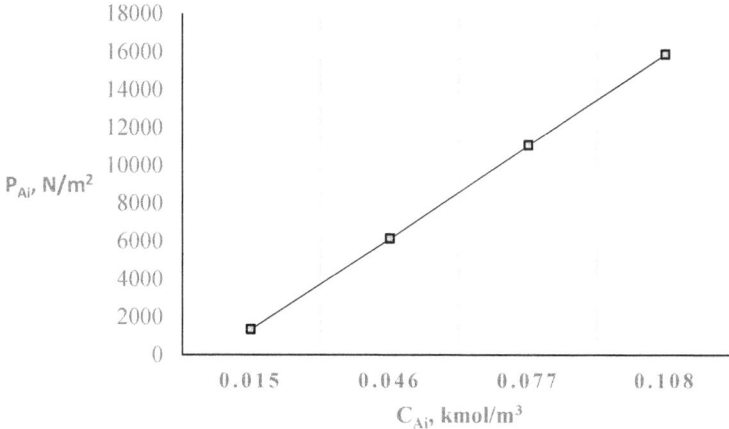

b) Evaluating the interfacial composition in both phases

The given equilibrium concentrations of SO_2 are given for the gas phase by

$$P_A = y_A P_T = 0.1 x 1.013 x 10^5 = 1.013 x 10^4, N/m^2 \qquad (9)$$

In the liquid phase, for the 0.4% solution, $\rho = 990$ kg/m³, $x = 0.004$ so that $C_A = 990 \; x \; 0.004/64 = 0.062$ kmol/m³. The values of P_A, C_A obtained define a point on a line, of slope, $- k_L/k_G$, which cuts the equilibrium line at the interfacial concentrations, P_{Ai}, C_{Ai}. The equation of this line is

$$\frac{1.013 x 10^4 - P_A}{0.062 - C_A} = -\frac{k_L}{k_G} \qquad (10)$$

from which

$$P_A = 1.013 x 10^4 + \frac{k_L}{k_G}(0.062 - C_A), \frac{m}{s} \cdot \frac{m^2.s}{kmol} \cdot \frac{N}{m^2} \cdot \frac{kmol}{m^3}$$

$$= 1.013 \; x \; 10^4 + \frac{2.1695 \; x \; 10^{-4}}{1.566 \; x \; 10^{-9}} \; (0.062 - C_A), \frac{N}{m^2}$$

$$= 18719.34 - 138537.68 C_A, \frac{N}{m^2} \qquad (11)$$

Plotting this operating line in the same graph as the equilibrium line gives the intersection of both lines from which the interfacial concentrations can be read off as $P_{Ai} = 9.451 \; x \; 10^3$ N/m^2 or 71.3 mm Hg and $C_{Ai} = 0.0669$ kmol/m^3 or 0.004 mole fraction.

Determining interfacial Pressure and Concentration in Example 3.7

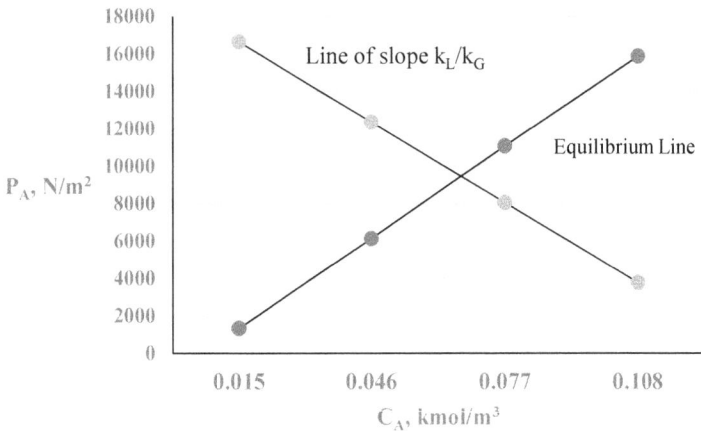

Example 3.8

Estimate the film mass transfer coefficient, k_G, for the flow of a dilute gas stream of air and SO$_2$ at a velocity of 3.05 m/s over a wet, plane, surface. In a previous heat transfer experiment over a geometrically similar surface, a heat transfer coefficient of 64.2 W/m^2.K was determined at a gas velocity of 6.10 m/s. Take the diffusivity of SO$_2$ through air, D, as 9.29 x 10^{-6} m/s, the specific heat capacity of air, Cp = 1.005 kJ/kg.K, thermal conductivity of air, k = 0.016 W/m.K, density of air, ρ = 1.28 kg/m^3 and the Prandtl number for air, Pr = 0.7. The Reynolds's analogy for mass, heat, and momentum, transfer is given as

$$\frac{k_G}{u} = \frac{h}{\rho Cp u} = \frac{R}{\rho u^2} \qquad (1)$$

and the Chilton - Colburn analogy as

$$j_H = \frac{h}{\rho Cp \, u}.Pr^{2/3} = 0.023 Re^{-0.2} \tag{2}$$

for heat and momentum transfer;

$$j_D = \frac{k_G}{u}.Sc^{2/3} = 0.023 Re^{-0.2} \tag{3}$$

for mass and momentum transfer. The symbols have their usual meanings.

Answer

The diagram below illustrates the cross section of the system. The mass transfer coefficient will be estimated using both the Reynold's and the Chilton - Colburn analogies.

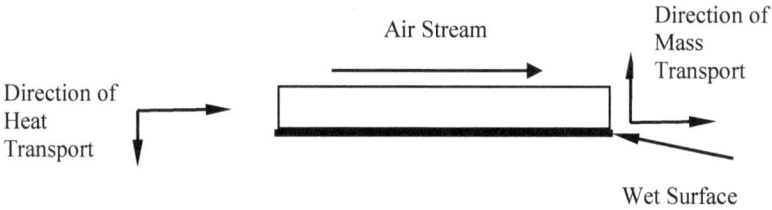

Using Reynolds's Analogy

Since the air is dilute, using the physical properties of air only is not likely to introduce serious errors in the calculation. Hence with u = 6.10 m/s, h = 64.2 W/m^2.K, we can evaluate each term in the analogy. The friction factor is calculated to be, from (1)

$$\frac{R}{\rho u^2} = \frac{h}{\rho Cp \, u} = \frac{64.2}{1.28 x 1.005 x 1000 x 6.10}, \frac{W}{m^2.K}.\frac{m^3}{kg}.\frac{kg.K}{kJ}.\frac{kJ}{J}.\frac{s}{m}$$

$$= 0.00818 \quad \text{(dimensionless)} \tag{4}$$

The dimensionless mass transfer coefficient can be evaluated by making use of the knowledge that the friction factor is not very sensitive to velocity variations, especially, in turbulent flow. Thus the friction factor at u = 6.10 m/s is likely to be the same as that for u = 3.05 m/s. That is

$$\frac{k_G}{u} = \frac{R}{\rho u^2} = 0.00818 \tag{5}$$

from which

$$k_G = 0.00818 x u = 0.00818 x 3.05, \, m/s = 0.025 m/s \quad Ans.$$

Using the Chilton - Colburn Analogy

Since the j-factor for heat transfer, j_H, and the j-factor for mass transfer, j_D, are, each, equal to $0.023\text{Re}^{-0.2}$, from (2) and (3)

$$j_H = j_D = \frac{h}{\rho \, Cp \, u}.\text{Pr}^{2/3} = \frac{64.2 \, x \, (0.7)^{2/3}}{1.28 \, x \, 1005 \, x \, 6.10}.\frac{W}{m^2.K}.\frac{m^3}{kg}.\frac{kg.K}{J}.\frac{s}{m}$$

$$= 0.00645 \quad \text{(dimensionless)} \tag{6}$$

Using the expression for j_D (equation (3)) at $u = 3.05$ m/s, we get

$$j_D = \frac{k_G}{u}.Sc^{2/3} \quad \text{or} \quad k_G = \frac{j_D \cdot u}{Sc^{2/3}} = \frac{0.00645 \, x \, 3.05}{Sc^{2/3}} \tag{7}$$

To evaluate the Schmidt number, Sc, we use the Prandtl number and the data given to get $\text{Pr} = Cp \, \mu / k$ or

$$\mu = \frac{k.\text{Pr}}{Cp} = \frac{0.016 \, x \, 0.7}{1005}, \frac{W}{m.K}.\frac{kg.K}{J} = 1.1144 \, x \, 10^{-5}, \frac{kg}{m.s} \tag{8}$$

so that

$$Sc = \frac{\mu}{\rho \, D} = \frac{1.1144 \, x \, 10^{-5}}{1.28 \, x \, 9.29 \, x \, 10^{-6}}, \frac{kg}{m.s}.\frac{m^3}{kg}.\frac{s}{m^2} = 0.94 \tag{9}$$

Substituting this value of the Schmidt number in equation (7), we get that

$$k_G = \frac{0.00645 \, x \, 3.05}{Sc^{2/3}} = \frac{0.00645 \, x \, 3.05}{(0.94)^{2/3}} = 0.0205\frac{m}{s} \quad Ans.$$

Note that the Reynolds's analogy, in this problem, overestimates k_G by about 22% compared to the Chilton - Colburn analogy.

Example 3.9

Mass transfer, in a turbulent boundary layer formed over a flat plate, was defined in terms of the local Sherwood number, Sh_x, by

$$Sh_x = 0.03 \text{Re}_x^{4/5} \, Sc^{1/3} \tag{1}$$

where Re_x is the Reynolds's number at a distance, x, from the leading edge, Sc the Schmidt number and Sh_x the Sherwood number at the same distance x from the leading edge, all given as

$$\text{Re}_x = \frac{\rho x u}{\mu}; \quad Sc = \frac{\mu}{\rho D}; \quad Sh_x = \frac{x k_G}{D} \tag{2}$$

where μ = viscosity of the fluid, ρ = density of the fluid, D = diffusivity of solute in the fluid, and k_G the mass transfer coefficient. If Re = 4000, Sc = 0.7 and D_{AB} = 1.8 x 10^{-9} m²/s, calculate

a). the mass transfer coefficient at a distance 0.6m from the leading edge

b). the j_D factor for mass transfer under these conditions.

Answer

a).   Evaluating the mass transfer coefficient

From the expression for the Sherwood number given in (1) and (2)

$$Sh_x = \frac{x k_G}{D} = 0.03 \text{Re}_x^{4/5} Sc^{1/3} \tag{3}$$

This gives

$$k_G = 0.03 \frac{D}{x} \text{Re}_x^{4/5} Sc^{1/3} = 0.03 x \frac{1.18 x 10^{-9}}{0.6} x (4000)^{4/5} x (0.7)^{1/3}, \frac{m^2}{s}.\frac{1}{m}$$

$$= 3.989 x 10^{-8} \; m/s \quad Ans.$$

b).   Evaluating the j_D factor

The j_D factor for mass transfer is given by

$$j_D = \frac{k_G}{u} . Sc^{2/3} \tag{4}$$

but we do not have a value for u. Since $\text{Re}_x = \frac{\rho x u}{\mu}$ and $Sc = \frac{\mu}{\rho D}$ we

can get   that $\text{Re} . Sc = \frac{\rho x u}{\mu} . \frac{\mu}{\rho D} = \frac{x u}{D}$ from which we get that

$u = \frac{D}{x}.\text{Re}.Sc$. That is

$$u = \frac{D}{x}.\text{Re}.Sc = \frac{1.18 x 10^{-9}}{0.6} x \; 4000 x \; 0.7, \frac{m^2}{s}.\frac{1}{m} = 5.507 x 10^{-6}, \frac{m}{s} \tag{5}$$

Substituting (5) and the value of k_G obtained in part (a) of this example into (4), we get that

$$j_D = \frac{k_G}{u}.Sc^{2/3} = \frac{3.989x10^{-8}}{5.507x10^{-6}} x(0.7)^{2/3}, \frac{m}{s}.\frac{s}{m} = 5.711x10^{-3} \quad Ans.$$

Example 3.10

Calderbank (1967) suggested the following equation for mass transfer in bubbles formed in agitated electrolytes

$$k_L = 0.31 \left(\frac{\Delta \rho \mu_L g}{\rho_L^2} \right)^{1/3} Sc^{-2/3} \qquad (1)$$

where k_L = mass transfer coefficient; μ_L = viscosity of the liquid, ρ_L = density of the liquid; $\Delta\rho$ = density difference between liquid and gas = $(\rho_L - \rho_G)$; Sc = Schmidt number = $\mu/\rho D$; μ_0 = viscosity of the gas; g = acceleration due to gravity; D_G = diffusivity of the gas in the liquid. Evaluate this equation for the case of CO_2 absorption in water at 300 K and 1 atm. total pressure.

Answer

The first step is to consult data tables in the literature to obtain experimentally determined values.

Thus, from Welty (1978), Appendix A2 and A3, we get,

for water at 300 K	ρ_L = 996 kg/m^3	μ_L= 0.86 x 10^{-6} Ns/m^2
for CO_2	ρ_G = 1.7967 kg/m^3	μ_G= 1.4948 x 10^{-7}, Ns/m^2

From Treybal (1980), Table 2.4, the closest value of D for CO_2 is given at 293 K as 1.77 x 10^{-9}, m^2/s This is at a temperature 7 K short of 300 K. Let us try an estimate of D using the Wilke and Chang (1955) correlation for dilute solutions of non-electrolytes.

$$D_{AB} = \frac{1.173x10^{-13}(\varphi M)^{0.5} T}{\mu v_M^{0.6}} \qquad (2)$$

where

D_{AB} = diffusivity of gas, A, in liquid, B, m^2/s

φ = association factor for the solvent = 2.6 for water (2.26 also recommended)

M = molecular weight of solvent

56

$\mu =$ viscosity of solvent, mNs/m^2

$v_M =$ molar volume of solvent at its boiling point =
0.0340 m^3/kmol for CO_2

$T =$ degrees Kelvin

Hence

$$D_{AB} = \frac{1.173x10^{-13}(2.6x18)^{0.5}x300}{0.86x10^{-3}(0.0340)^{0.6}} = 2.129x10^{-6}, \frac{m^2}{s} \qquad (3)$$

This value is several orders of magnitude higher than the experimental value at only 7 K short of 300 K. Hence, we shall use the experimental value from Treybal (1980) in the Calderbank equation. Thus

$$Sc = \frac{\mu}{\rho D} = \frac{1.4948x10^{-7}}{1.7967x1.77x10^{-9}} = 47 \qquad (4)$$

$$\Delta\rho = \rho_L - \rho_G = 996 - 1.7967 kg/m^3 = 994.2\, kg/m^3 \qquad (5)$$

Substituting (4), (5) and given values in (1)

$$k_L = 0.31\left(\frac{\Delta\rho\,\mu_L\,g}{\rho_L^2}\right)^{1/3} Sc^{-2/3}$$

$$= 0.31\left(\frac{994.2\,x\,0.86x10^{-6}x9.81}{(996)^2}\right)^{1/3} x(47)^{-2/3}, \left(\frac{kg}{m^3}\cdot\frac{kg}{m.s}\cdot\frac{m}{s^2}\cdot\frac{m^6}{kg^2}\right)^{1/3}$$

$$= 1.564x10^{-4}\, m/s \qquad Ans$$

References For Chapter Three

1. Calderbank (October, 1967); Chemical Engineer No. 212; Brit. Inst. Chem. Engrs., London

2. Treybal R. E., Mass Transfer Operations; Chapter 3; McGraw Hill Book Company, N. Y., USA, 1980.

3. Welty J. R., Engineering Heat Transfer, SI Version; John Wiley & Sons, N.Y., USA, 1978.

CHAPTER FOUR:
MASS TRANSFER OPERATIONS

A mass transfer operation, as stated earlier, is a deliberate action, by humans, to exploit mass transfer phenomena or process for economic or other gain. Most mass transfer operations are motivated by the need to make maximum contact between the phases in which mass transfer is expected to occur. This implies, since mass transfer occurs only across an interface, the employment of means to increase interfacial contact between the phases. Many systems have been proposed and tested but the most popular ones include:

i.   The Packed Bed Column

This is a column in which the heavy (liquid) phase flows down and over a bed of discrete solids through which the lighter (gas) phase flows in an upward direction. The idea is that the bed of discrete solids breaks up the continuous, heavy and lighter phases into disperse phases. These disperse phases may be made up of thin films, spherical drops or bubbles which have greater interfacial surface area per unit volume than the continuous phase. Its hydrodynamics, types of packing, advantages and disadvantages are treated elsewhere (Treybal, 1980). The objective is to determine, for a given type of packing, of which there may be several, the height of packing required to do the specified mass transfer duty.

The original development of the packed bed was the fixed, packed bed in which the packing was stationary and both the heavy and lighter phases flowed through it. This later gave rise to the moving bed in which the solids were transported along a channel or conduit through which the light or heavy phase or both were flowing. Mixing was not, particularly, due to either of the phases.

Later on, the fluidised bed was developed in which the lighter phase was used to achieve uniform and continuous stirring of the solids which, either took part in the mass transfer, often with chemical reaction, or was inert and, simply, provided uniform stirring and dispersal of both phases. The most current development involves the bulk rotation of the fixed bed such that, not only mass transfer takes place, but also chromatographic separation.

ii.   The Staged Plate Column

In this type of column, the heavy and light phases flow, counter-currently, over perforated trays (plates) or in agitated stages such that maximum surface area and mass transfer are obtained in each tray or stage. The interest here is to determine the number of plates or stages required to effect a given separation or vice versa.

iii.   The Direct Contact System

This consists of a system or device in which the lighter phase is bubbled through the heavy phase thereby causing both system agitation and increase in interfacial area.

In all cases, the basic premise is that, at the given temperature and pressure, there is a series, of maximum solubility points, for every concentration in one phase, which is in equilibrium with every other concentration in the other phase. Such maximum solubility points define the equilibrium curve of the system and give the expected driving force for mass transfer between what is possible (the equilibrium position) and what is desired (the operating position). The knowledge of these two positions makes it possible to determine either the height of packing or the number of stages required to effect a given mass transfer.

This chapter solves eleven problems which illustrate

i.   The various ways of plotting equilibrium data once these have been obtained from suitable experiments

ii. The method of estimating the number of theoretical stages using both a graphical and a calculation method. The actual number of stages is obtained by dividing the theoretical number of stages by either the stage or column efficiency. These efficiencies range between 30% and 70% with the mean around 50%.

iii. The methods of estimating the height of packing in a packed tower.

Example 4.1

The equilibrium data for the Benzene-Wash Oil system are given as

Plot the equilibrium data in terms of mole fractions.

X, moles Benzene/mole of wash oil	0.02	0.04	0.06	0.08	0.10	0.12	0.14
Y, moles Benzene/mole steam	0.07	0.14	0.22	0.31	0.405	0.515	0.650

Plot these equilibrium data in terms of mole fractions.

Answer

The data are given in mole ratios, X and Y. To convert to mole fractions, x and y, use

$$x = \frac{X}{1+X}, \quad y = \frac{Y}{1+Y} \tag{1}$$

to obtain

x	0.0196	0.0385	0.0566	0.0741	0.0909	0.1071	0.1228
y	0.0654	0.1228	0.1803	0.2366	0.2883	0.3399	0.3939

These points are plotted in Figure 4.1 below.

Fig. 4.1: Plot of Equilibrium Data in terms of Mole Fractions

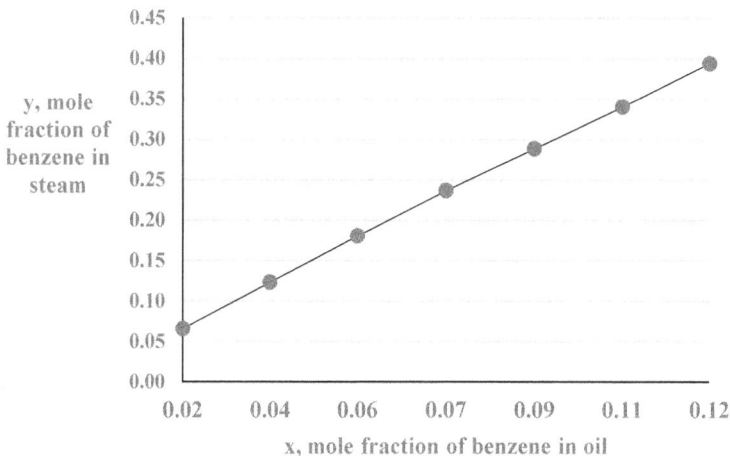

y, mole fraction of benzene in steam

x, mole fraction of benzene in oil

Example 4.2

The equilibrium partial pressure of water vapour, in contact with a certain silica gel, on which water is absorbed, is given, at 25 C, as follows:

Partial pressure H_2O, mm Hg	0	2.14	4.74	7.13	9.05	10.9	12.6	14.3	16.7
Kg H_2O/100 kg dry gel	0	5	10	15	20	25	30	35	40

Plot the equilibrium data using X = kmol H_2O/kg dry gel and Y = kmol H_2O/kmol dry air. The total pressure is 1 atm.

Answer

To convert kg H_2O/100kg dry gel to X = kmol H_2O/kg dry gel, we must understand that kg H_2O/l00kg dry gel = 100 times the kg H_2O/kg dry gel so that kg H_2O/kg dry gel = 1/100 x kg H_2O/100kg dry gel.

To convert kg H_2O/kg dry gel to kmol H_2O/kg dry gel, we simply divide it by the molecular weight of H_2O. Thus

$$X, \frac{kmol\, H_2O}{kg\, dry\, gel} = \frac{1}{100} x \frac{kg\, H_2O}{100 kg\, dry\, gel} x \frac{kmol\, H_2O}{18 kg\, H_2O} \qquad (1)$$

To convert partial pressure to mole fraction, y, we remember that

$$y = \frac{Partial\, pressure, mmHg}{Total\, pressure, mmHg} = \frac{Partial\, pressure, mmHg}{760, mmHg} \qquad (2)$$

To convert mole fraction, y, to mole ratio, Y, we remember that

$$y = \frac{Y}{1+Y} \quad or \quad Y = \frac{y}{1-y} \qquad (3)$$

Applying (1), (2) and (3) to the data given, we obtain the Table below

kg $_2$O/100 kg dry gel	Partial pressure, H_2O, mm Hg	Mole fraction, y, H_2O in air	kmol H_2O/ kg dry gel, X	kmol H_2O/ kmol dry air, Y
0	0	0	0	0
5	2.14	0.002816	0.002778	0.002824
10	4.74	0.006237	0.005556	0.006276
15	7.13	0.009382	0.008333	0.00947
20	9.05	0.011908	0.011111	0.012051
25	10.9	0.014342	0.013889	0.014551
30	12.6	0.016579	0.016667	0.016858
35	14.3	0.018816	0.019444	0.019177
40	16.7	0.021974	0.022222	0.022467

This is plotted in Fig. 4.2 below.

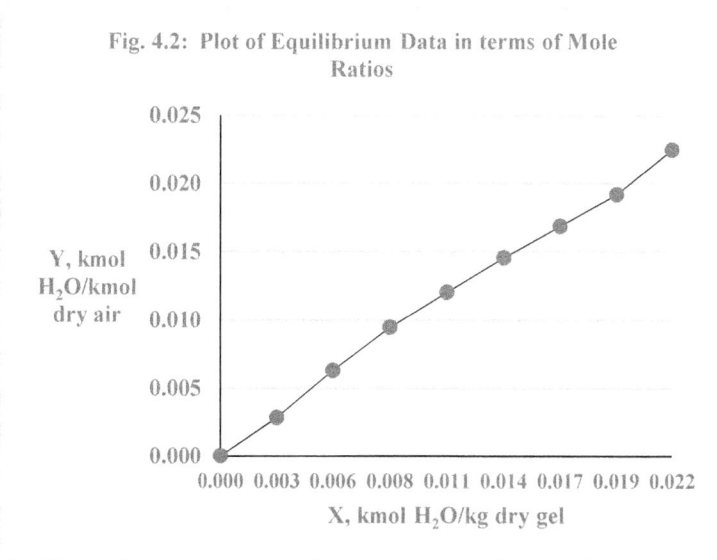

Fig. 4.2: Plot of Equilibrium Data in terms of Mole Ratios

Example 4.3

The equilibrium data for the SO_2-Water system are given as follows:

X	0.02	0.05	0.1	0.3	0.5	0.7	1.0	1.5	2.5	5.0
P_{SO_2}	0.5	1.2	3.2	14.1	26	39	59	92	161	336

where X = mass of SO_2 /100 Mass of H_2O and P_{SO_2} = partial pressure of

SO$_2$ in air, mm. Hg. The total pressure is 2 atm. Plot the equilibrium data in mole fraction units. You are given (can find out) that 1 atm. = 760 mm. Hg; the molecular weights of SO$_2$ = 64, H$_2$O = 18, O$_2$ = 32, N$_2$ = 28. The amount of O$_2$ in air is 21 % while that of N$_2$ in air is 79% by volume.

Answer

In the liquid phase, the Mass of SO$_2$/100 Mass of H$_2$O = 100 times the mass ratio of SO$_2$ to H$_2$O = $100X^1$, where X^1 is the mass ratio of SO$_2$ to H$_2$O. Thus

$$X^1, \frac{Mass\ SO_2}{Mass\ H_2O} = \frac{1}{100} x \frac{Mass\ SO_2}{100\ Mass\ H_2O} = \frac{X}{100} \quad (1)$$

To convert mass ratio, X^1, to mole ratio, X_m, we remember that

$$X_m = \frac{(mass\ SO_2)/64.}{(mass\ H_2O)/18} = \frac{18\ X^1}{64} \quad (2)$$

Mole fraction, x, is then, in terms of the original data given,

$$x = \frac{X_m}{1+X_m} = \frac{18/64.X^1}{1+18/64.X^1} = \frac{18\ X^1}{64+18\ X^1}$$

$$= \frac{18\dfrac{X}{100}}{64+18\dfrac{X}{100}} = \frac{0.18X}{64+0.18X} \quad (3)$$

To convert partial pressure to mole fraction, y, in the gas phase, we remember that

$$y = \frac{Partial\ pressure, mm\ Hg}{Total\ pressure, mm\ Hg} = \frac{Partial\ pressure, mm\ Hg}{2\ atm.\ x\ 760, mm\ Hg\ per\ atm.} \quad (4)$$

Applying (1), (2), (3) and (4) to the data given, we obtain the Table below

pSO2, mm Hg	Mass of SO2/100 Mass of H2O	Mass of SO2/Mass of H2O, x 10-2	Moles of SO2/Moles of H2O, x 10-4	Mole fraction SO2 in H2O, x, x 10-3	Mole fraction SO2 in Air, y, x 10-3
0.5	0.02	0.02	0.563	0.056	0.33
1.2	0.05	0.05	1.406	0.141	0.79
3.2	0.10	0.10	2.813	0.281	2.105
14.1	0.30	0.30	8.438	0.843	9.276
26.0	0.50	0.50	14.060	1.404	17.110
39.0	0.70	0.70	19.690	1.965	25.660
59.0	1.00	1.00	28.130	2.805	38.810
92.0	1.50	1.50	42.190	4.201	60.530
161.0	2.50	2.50	70.310	6.982	105.920
336.0	5.00	5.00	140.600	13.870	221.050

The x and y values in columns 5 and 6 are plotted in the Figure 4.3 below.

Fig. 4.3: Equilibrium Diagram for the SO_2-H_2O System

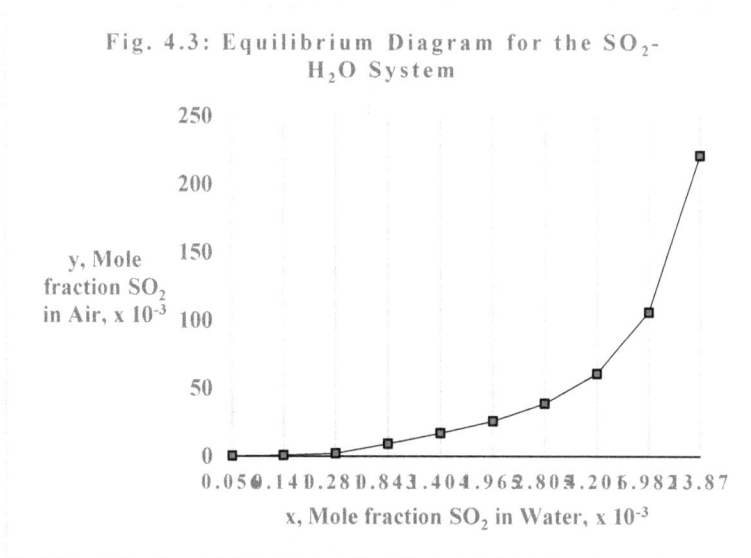

y, Mole fraction SO_2 in Air, x 10^{-3}

x, Mole fraction SO_2 in Water, x 10^{-3}

Example 4.4

A CS_2 - N_2 mixture is to be scrubbed in a counter-current manner with a hydrocarbon in order to recover the CS_2, which is a pollutant. The data for the process are as follows:

65

Inlet gas stream $Y_1 = 0.071$ kmol CS_2/kmol N_2; $G = 54.28$ kmol N_2/h

Exit gas stream $Y_2 = 0.005$ kmol CS_2/kmol N_2

Inlet oil stream $X_2 = 0$

Exit oil stream $X_1 = 0.113$ kmol CS_2/kmol oil

The equilibrium curve is given by

$$\frac{Y_e}{X_e} = 0.455\frac{1+Y_e}{1+X_e} \tag{1}$$

Plot the equilibrium and operating lines, indicating the entry and exit conditions,

Answer

The process can be represented, schematically, as follows:

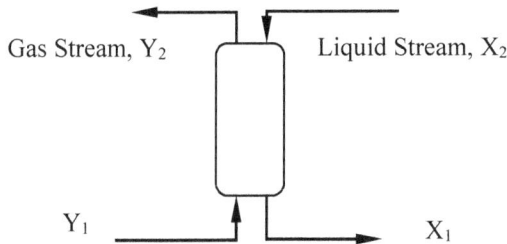

Gas Stream, Y_2 Liquid Stream, X_2

Y_1 X_1

To plot the equilibrium line, we rearrange the given equation and obtain

$$\frac{Y_e}{1+Y_e} = \frac{0.455X_e}{1+X_e} = m \quad let\ us\ say \tag{2}$$

Then $$Y_e = \frac{m}{1-m} = \frac{0.455X_e}{1+X_e}\cdot\frac{1+X_e}{1+0.545X_e} = \frac{0.455X_e}{1+0.545X_e} \tag{3}$$

By using values of $X_e = 0$ to $X_e = 0.16$, we can obtain, from the above, the following Table

X_e	0	0.02	0.04	0.06	0.08	0.10	0.12	0.14	0.16
Y_e	0	0.009	0.018	0.026	0.035	0.043	0.051	0.059	0.067

The operating line is obtained by a mass balance as follows.

$$G(Y_1 - Y_2) = L(X_1 - X_2) \qquad (4)$$

where G and L are the hourly, molar flow rate of nitrogen and the hydrocarbon oil, respectively. Equation (4) gives the value of L as

$$L = \frac{G(Y_1 - Y_2)}{(X_1 - X_2)} = \frac{54.28(0.071 - 0.005)}{(0.113 - 0)} = 31.7 \frac{kmol\,oil}{h} \qquad (5)$$

The generalised operating line equation can now be derived as

$$54.28(Y - 0.005) = 31.7(X - 0) \quad or \quad Y = 0.005 + 0.584X \qquad (6)$$

which can be tabulated as

X	Y
0	0.005
0.02	0.0167
0.04	0.0284
0.06	0.0400
0.08	0.0517
0.10	0.0634
0.12	0.0751
0.14	0.0868
0.16	0.0984

The equilibrium and operating values of X and Y are plotted in Figure 4.4 below showing gas and liquid inlet and exit compositions.

Fig. 4.4: Plot of Equilibrium and Operating Lines

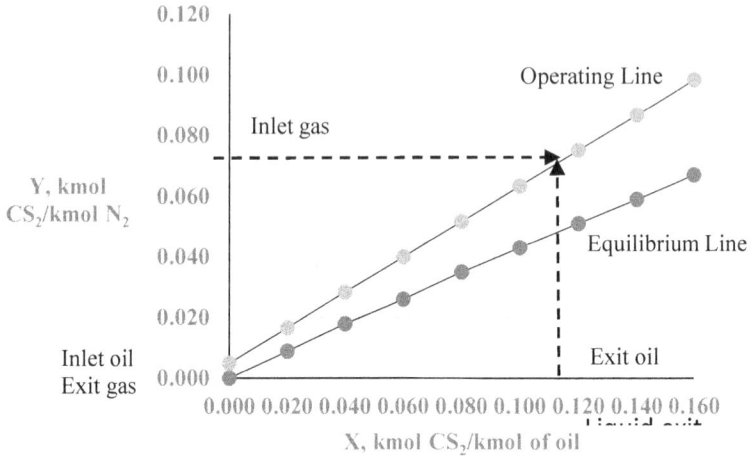

Example 4.5

If, in Example 4.4, the inert gas flow rate, G_S = 54.28 kmol N_2/h, the liquid /gas rate ratio, L/G = 0.584 and the Absorption Factor = 1.3, determine the number of theoretical plates required

    a.   graphically
    b.   using the Kremser-Souder-Brown equation

$$N = \frac{\log\left[\left(\dfrac{Y_{N+1} - m X_0}{Y_1 - m X_0}\right)\left(1 - \dfrac{1}{A}\right) + \dfrac{1}{A}\right]}{\log A} \tag{1}$$

where N = number of ideal plates and A = Absorption Factor.

Answer

    a)   <u>Graphical Solution</u>

The equilibrium and operating lines are drawn as in Example 4.4.

Since both the equilibrium and operating lines are, reasonably, straight, the number of ideal plates is marked out as shown in the Figure below.

Start from the exit condition. Draw a horizontal line from A, on the operating line, to point B on the equilibrium line. From B, on the equilibrium line, draw a vertical line to intercept the operating line at point C. Triangle ABC indicates one ideal plate. Similarly, CDE indicates the second ideal plate and so on until point K is reached. I J K should have been the last ideal plate. But point K is not quite the same as the inlet condition (0.113, 0.071). Hence, by similar triangles, the fraction of the ideal plate is given by KL/KM. Thus, graphically, the number of ideal plates = 5 + 3/24 = 5.125. In practice, the number is rounded up to 6 ideal plates or to 5 depending on the degree of overdesign desired and the accuracy of the data.

Graphical Determination of the Number of Ideal Plates in Example 4.5

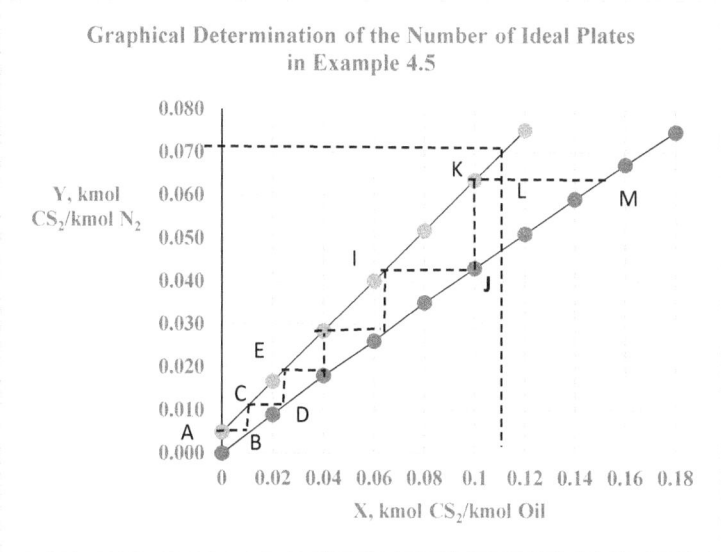

b)   Using the Kremser-Souder-Brown Equation

Since $X_2 = X_0 = 0$, the inlet liquid concentration, there is no need to evaluate m from $L/mG = A$, although in this problem, we know $L/G = 0.584$. Hence the Kremser-Souder-Brown equation reduces to

$$N = \frac{\log\left[\left(\frac{Y_2}{Y_1}\right)\left(1 - \frac{1}{A}\right) + \frac{1}{A}\right]}{\log A} = \frac{\log\left[\left(\frac{0.071}{0.005}\right)\left(1 - \frac{1}{1.3}\right) + \frac{1}{1.3}\right]}{\log 1.3} = 5.33$$

In practice, N is rounded up to 6 ideal stages. Note the small discrepancy between the results obtained graphically and from the Kremser-Souder-Brown equation.

Example 4.6

Benzene is to be absorbed from coal gas by means of a wash oil. The inlet gas contains 3% by volume of benzene and the exit gas should not contain more than 0.02% benzene by volume. The wash oil enters the tower solute free and at a circulation rate of 13.6 kg oil per 2.83 m³ of inlet gas, measured at STP. The operation is carried out at atmospheric pressure. The equilibrium data are

Benzene in oil, Wt. %	0.05	0.10	0.50	1.0	2.0	2.5	3.0
Equilibrium Partial Pressure of Benzene in Gas, mm. Hg	0.11	0.23	1.5	4.0	10	16	25

Determine, graphically, the number of transfer units (NTUs) in the tower in which differential contact between the oil and gas occurs (Coulson et al, 1980).

Answer

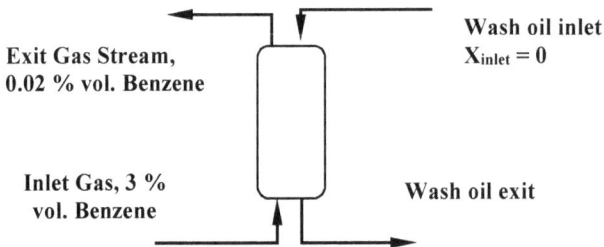

Since the liquid to gas ratio is given in terms of kg to m³, it is convenient to work in weight fractions in the liquid phase and in volume fractions in the gas phase. Thus, in the liquid phase,

$$Weight\ fraction = \frac{Weight\ per\ cent}{100} \qquad (1)$$

In the gas phase

$$Volume\ fraction = \frac{Partial\ pressure, mm\ Hg}{Total\ pressure, mm\ Hg} \qquad (2)$$

The equilibrium data are, then, expressed as follows:

Benzene in oil, Wt. Fraction, x_e	Benzene in gas, Vol. Fraction, y_e	Benzene in oil, Wt. Fraction, x_e	Benzene in gas, Vol. Fraction, y_e
0.0005	0.0002	0.02	0.0132
0.001	0.0003	0.025	0.0211
0.005	0.0020	0.03	0.0329
0.01	0.0053		

To obtain the operating line, we need to know the concentration of benzene at both terminal points of the tower. Since the concentration of benzene in the exit stream is not specified, it is obtained from a mass balance across the tower using the information given. This is obtained as follows:

For every $13.6(x_{exit} - x_{inlet})$kg benzene absorbed by the wash oil, $2.83(0.03 - 0.0002) = 0.084m^3$ of benzene are removed from the gas stream. This volume is equivalent, at STP, to $0.084 x\ 78/22.4$ kg benzene. Thus, we can say that $13.6(x_{exit} - 0) = 0.294$ or $x_{exit} = 0.0216$.

Thus, we know, now, that the terminal points of the operating line are points A (0.0216, 0.03) and B (0, 0.0002) for which the operating line is obtained from the material balance $2.83(y - 0.0002) = 13.6(x - 0)$ as

$$y = 0.0002 + 4.806x \qquad (3)$$

This equation is tabulated within the range of x of the operating line as

Benzene in oil, Wt. Fraction, x	Benzene in gas, Vol. Fraction, y	Benzene in oil, Wt. Fraction, x	Benzene in gas, Vol. Fraction, y
0.0005	0.0026	0.02	0.0963
0.001	0.0050	0.025	0.1204
0.005	0.0242	0.03	0.1444
0.01	0.0483		

The operating line is plotted along with the equilibrium line in Figure 4.6

below. The number of transfer units is determined as follows:

i.   At regular intervals on the x-axis, determine the mid-point of the vertical distance between the operating and equilibrium lines

ii.   Join these mid-points by the curve, EF

iii.   Starting from point B, on the operating line, draw a horizontal line bisected by the curve, EF.
From the end of this horizontal line, draw a vertical line to intersect the operating line. This gives one transfer unit.

iv.   Continue as from (iii) using this new intersection point, on the operating line, to obtain the next transfer unit.

From Figure 4.6, 3 and a fraction rounded up to 4 transfer units are obtained as the answer.

Fig. 4.6: Graphical Determination of the Number of Transfer Units

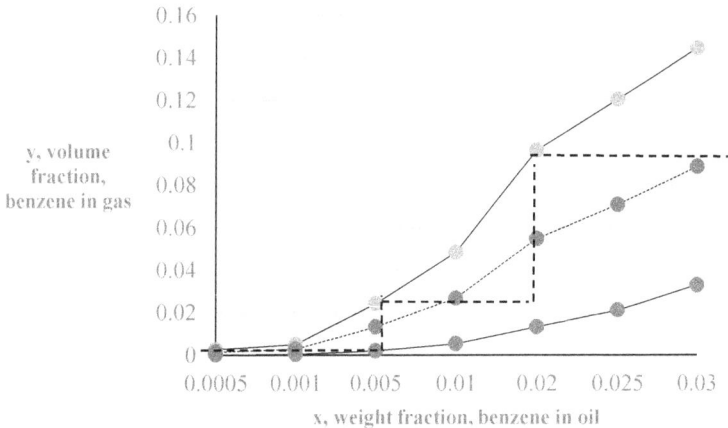

Example 4.7

A tower, 154 mm diameter and packed for 1270 mm, reduces the ammonia concentration from 4.5 to 1.3 volume percent in an air - ammonia stream, fed at the rate of 0.014 kg/s. An ammonia-free water stream flows, counter-currently, at the rate of 9.09×10^{-3} kg/s. The tower operates isothermally at 20 C and 1 atm. Under these conditions, the equilibrium data are as follows:

X_e, kmol NH_3/kmol H_2O	0.0164	0.0252	0.0349	0.0445	0.0722
Y_e, kmol NH_3/kmol Air	0.021	0.032	0.042	0.053	0.080

Determine a) the number of transfer units (NTU)
 b) the gas based overall mass transfer coefficient, K_ya
Take N = 14, H = 1, O = 16. Air is 21% oxygen and 79% nitrogen.

Answer

The tower may be represented, schematically, as follows

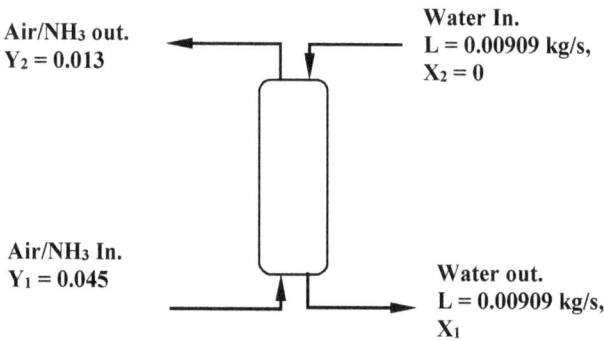

Air/NH₃ out.
$Y_2 = 0.013$

Water In.
L = 0.00909 kg/s,
$X_2 = 0$

Air/NH₃ In.
$Y_1 = 0.045$

Water out.
L = 0.00909 kg/s,
X_1

Since the equilibrium data is given in mole ratios, the calculation will be done in those terms.

i. First of all, we determine the molar rate of flow of carrier gas (air), G_S, and liquid (water), L_S.

The mean molecular weight of air

$$= 0.21\,0.21x32 + 0.79x28 = 28.84, \frac{kg}{kmol} \quad (1)$$

The mean molecular weight of the liquid stream, water, = 18.

The molar flow rate of air that is free of ammonia

$$= \frac{0.014}{28.84}(1 - 0.045), \frac{kg}{s} . \frac{kmol}{kg} = 4.636x10^{-4}, \frac{kmol}{s} \quad (2)$$

The tower cross-section, A_S, is

$$\frac{\pi d^2}{4} = \frac{\pi x (0.154)^2}{4} = 0.01863, m^2 \quad (3)$$

Molar flow rate of air, G_S, based on the tower cross-section, A_S, is then

$$G_S = \frac{4.636x10^{-4}}{0.01863}, \frac{kmol}{s} \cdot \frac{1}{m^2} = 0.0249, \frac{kmol}{m^2.s} \tag{4}$$

Molar flow rate of the water stream, L_S, based on tower cross-section, A_S, is also

$$L_S = \frac{9.09x10^{-3}}{18x0.01863}, \frac{kg}{s} \cdot \frac{kmol}{kg} \cdot \frac{1}{m^2} = 0.0271, \frac{kmol}{m^2.s} \tag{5}$$

ii. Then we determine the operating line and terminal conditions by a mass balance

$$G_S \left(Y_1 - Y_2 \right) = L_S \left(X_1 - X_2 \right) \tag{6}$$

Since $Y = \dfrac{y}{1-y}$ and $X = \dfrac{x}{1-x}$ so that $Y_1 = \dfrac{0.045}{1-0.045} = 0.047$,

$Y_2 = \dfrac{0.013}{1-0.013} = 0.013$ and $X_2 = 0$, the mass balance, equation (6),

becomes $0.0249(0.047 - 0.013) = 0.0271(X_1 - 0)$ from which

$$X_1 = \frac{0.0249x0.034}{0.0271} = 0.031 \tag{7}$$

The operating line is, thus, defined by the points $(X_1, Y_1) = (0.031, 0.047)$ and $(X_2, Y_2) = (0, 0.013)$. Using a general X and Y in the mass balance equation

$$0.0249(Y - 0.013) = 0.0271(X - 0) \qquad\qquad \textit{from (6)}$$

the operating line is obtained as

$$Y = 0.013 + 1.0884X \tag{8}$$

This is plotted with the equilibrium line in Figure 4.7 below.

iii. The equilibrium curve is, also plotted directly in Figure 4.7, since the equilibrium data are, already, in desirable form, that is, in mole ratios.

iv. We can, now, determine the number of transfer units, graphically, from Fig. 4.7, as $\underline{NTU = 3 + 14/23 = 3.61}$ Ans.

Fig. 4.7: Determining the NTU in a Packed Tower

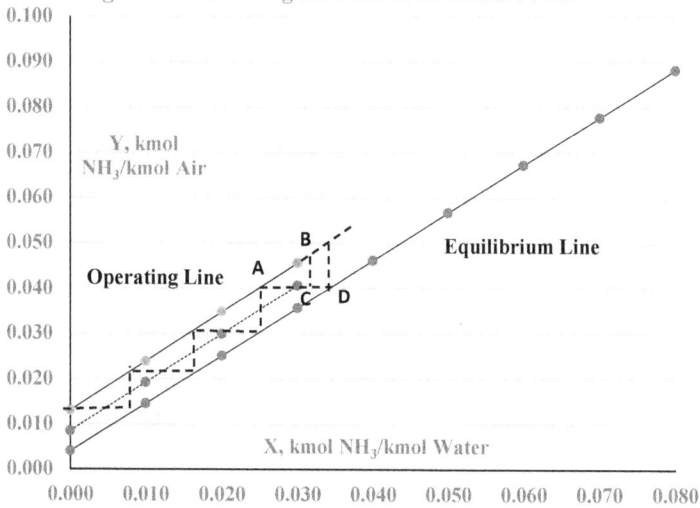

v. Since the height of packing, $Z = HTU \times NTU$, then HTU (height of a transfer unit) is given by

$$HTU = \frac{Z}{NTU} = \frac{1.27}{3.61}, \frac{m}{mole\ ratio} = 0.351\frac{m}{mole\ ratio} \quad (9)$$

v. Since $HTU = G_S/K_ya$, K_ya, the overall mass transfer coefficient, based on the gas phase, is then equal to

$$K_ya = \frac{G_S}{HTU} = \frac{0.0249}{0.351}, \frac{kmol}{s.m^2}.\frac{1}{m} = 0.0709\ \frac{kmol}{s.m^3}\ per\ mole\ ratio$$

Example 4.8

The equilibrium data for SO_2 and water at 293 K (Coulson and Richardson, Chem. Eng. Vol. II) are

kmol SO_2 /1000 kmol H_2O	0.056	0.14	0.28	0.42	0.56	0.84	1.405
kmol SO_2/ 1000 kmol inert gas	0.7	1.6	4.3	7.9	11.6	19.4	36.3

Determine the number of transfer units for an absorption tower through which a smelter gas, containing 3.5%, by volume, SO_2 is passed so that the exit gas has SO_2 at a partial pressure of 1.14 kN/m^2. Water enters the tower, solute free, at the rate of 43 kmol/s and leaves with 1.145×10^{-3}

kmol SO_2 per kmol water. Assume that the tower operates at 1 atm. (1.013×10^5, N/m^2).

Answer

The absorption tower may be represented, schematically, as follows

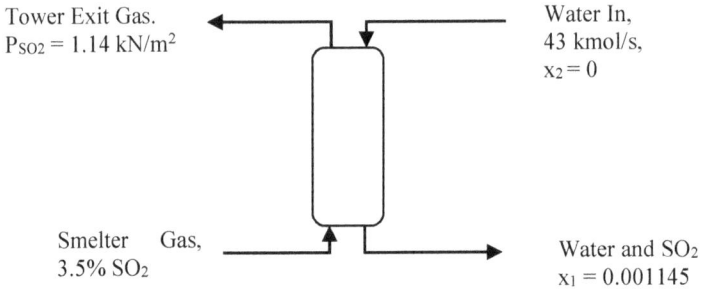

Tower Exit Gas.
$P_{SO2} = 1.14$ kN/m^2

Water In,
43 kmol/s,
$x_2 = 0$

Smelter Gas,
3.5% SO_2

Water and SO_2
$x_1 = 0.001145$

i.   Determine the terminal conditions

Smelter gas inlet: Since mole fraction of SO_2, $y_1 = 0.035$, the mole ratio of SO_2 to gas is

$$Y_1 = \frac{0.035}{1 - 0.035} = 0.036 \qquad (1)$$

Smelter gas exit: the mole fraction is given by

$$y_2 = \frac{partial\ pressure\ of\ SO_2}{Total\ pressure} = \frac{1.14}{101.3} = 0.011 \qquad (2)$$

The mole ratio is, then

$$Y_2 = \frac{0.011}{1 - 0.011} = 0.011 \qquad (3)$$

Water inlet: mole ratio SO_2 to water, $X_2 = 0$.
Water exit: mole ratio SO_2 to water, $X_1 = 1.145 \times 10^{-3}$.

Plot the equilibrium and operating lines, using the equilibrium data given and the (0.001145, 0.036) and (0, 0.011) calculated above. The molar flow rate of inert gas, G, is obtained from an overall mass balance.

$$G(0.036 - 0.011), \frac{kmol\ inert\ gas}{s} \cdot \frac{kmol\ SO_2}{kmol\ inert\ gas}$$

$$= 43(0.001145 - 0), \frac{kmol\,water}{s} \cdot \frac{kmol\,SO_2}{kmol\,water} \qquad (4)$$

This gives us G as 1.9694 kmol/s. The general operating line equation then becomes

$$1.9694(Y - 0.011) = 43(X - 0) \quad \text{or} \quad Y = 0.011 + 21.834X \qquad (5)$$

Because kmol SO_2 / kmol = (1/1000) of kmol SO_2 /1000 kmol, the given equilibrium data is retabulated as shown in the Table below.

kmol SO_2 / kmol H_2O, x 10^{-4}	0.56	1.4	2.8	4.2	5.6	8.4	14.05
kmol SO_2 / kmol inert gas, x 10^{-3}	0.7	1.6	4.3	7.9	11.6	19.4	36.3

The plot of equation (5) and the given equilibrium data is shown in Figure 4.8a.

Determine the NTU, graphically, from Figure 4.8a. This found to be 3 + 10/15 = 3.67. Alternatively, the NTU may be determined using graphical integration. The area under the plot of $\dfrac{1}{Y - Y_e}$ versus Y gives the NTU.

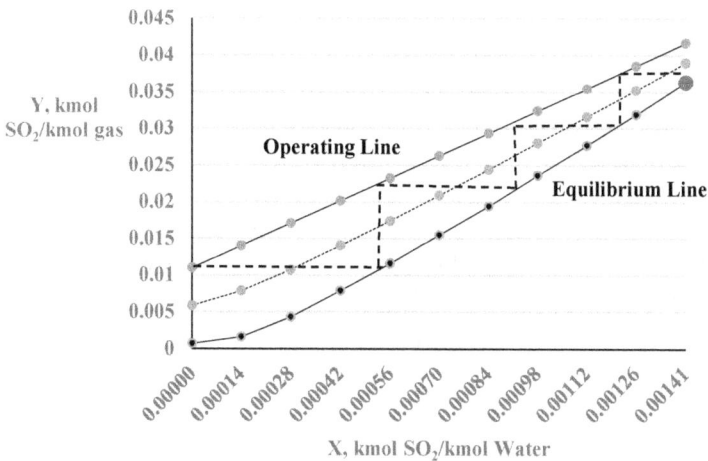

Fig. 4.8a: Determining the NTU by a geometric Method

The following values are obtained from either Figure 4.8a or from equation (5) and the equilibrium data and plotted in Figure 4.8b.

Y	$1/(Y-Y_e)$	
0.0110	97.087	f_0
0.0141	80.278	f_1
0.0171	70.043	f_2
0.0202	81.498	f_3
0.0232	86.006	f_4
0.0263	92.732	f_5
0.0293	100.598	f_6
0.0324	113.671	f_7
0.0355	128.964	f_8
0.0385	151.267	f_9
0.0417	185.985	f_{10}

Fig. 4.8b: Determining the NTU using a Numerical (Graphical) Integration Method

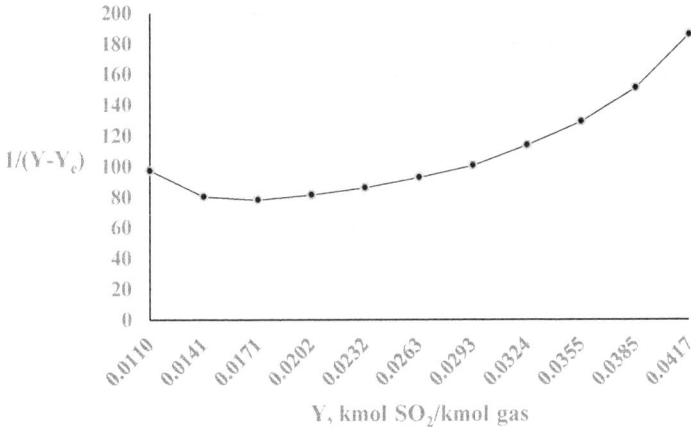

Simpson's numerical integration rule for two elements with ordinates f_0, f_1 and f_2 is given by

$$NTU = \int_{Y_1}^{Y_2} \frac{dY}{y - Y_e} = \frac{h}{3}\{f_0 + 4f_1 + f_2\}$$

For n even numbered elements, the composite Simpson's rule is given by (note that there are other specialist versions of Simpson's rule)

$$NTU = \int_{Y_1}^{Y_2} \frac{dY}{y - Y_e}$$

$$= \frac{h}{3}\{f_0 + 4f_1 + 2f_2 + 4f_3 + 2f_4 + 4f_5 + 2f_6 + \cdots \ldots 4f_{n-1} + f_n\}$$

This is shown to be true as follows

$$NTU = \int_{Y_1}^{Y_2} \frac{dY}{y - Y_e}$$

$$= \frac{h}{3}\{(f_0 + 4f_1 + f_2) + (f_2 + 4f_3 + f_4) + (f_4 + 4f_5 + f_6) + (f_6 + 4f_7 + f_8) + (f_8 + 4f_9 + f_{10})\}$$

$$= \frac{h}{3}\{(f_0 + 4f_1 + 2f_2 + 4f_3 + 2f_4 + 4f_5 + 2f_6 + 4f_7 + 2f_8 + 4f_9 + f_{10})\}$$

Hence

$$NTU = \frac{0.00307}{3}\{(97.087 + 4 \ x \ 80.278 + 2 \ x \ 70.043 + 4 \ x \ 81.498$$
$$+ 2 \ x \ 86.006 + 4 \ x \ 92.732 + 2 \ x \ 100.598$$
$$+ 4 \ x \ 113.671 + 2 \ x \ 128.964 + \ 4 \ x \ 151.267$$
$$+ 185.985)\} = 3.205$$

Note the difference between the values obtained from the geometrical and the numerical (graphical) integration, procedures

Example 4.9

It is desired to estimate the height of a liquid phase transfer unit for the absorption of CO_2 in water at 30 C in a packed tower, 1.5m high, using Cornell's method which states that (Coulson, Richardson & Sinaott, 1980)

$$H_L = 0.305\Phi_H (Sc_L)^{0.5} K_3 \cdot \left(\frac{Z}{3.05}\right)^{0.15} \qquad (1)$$

where

H_L = height of a liquid phase transfer unit, m

Φ_H = correction factor = 0.064 for a liquid rate of 4.53 kg/m².s

Sc_L = Schmidt number for the liquid phase = $\mu_L/\rho_L D_L$

K_3 = 0.85, the 60% flooding correction factor

Z = height of column, m

μ_L = $0.01\,\rho_L^{0.5}$, mNs/m²

ρ_L = 1000 kg/m³

D_L $= \dfrac{1.173x10^{13}\,(\phi.M)^{0.5}\,T}{\mu_S.V_M^{0.6}}$

ϕ = association factor for the solvent = 2.6 for water

μ_S = viscosity of solvent, mNs/m²

T = temperature, deg. K

V_M = molar volume of solute at its boiling point, m³/kmol;
= 0.0340 for CO_2

Answer

Since

$$\Phi_H = 0.064$$
$$K_3 = 0.85$$
$$Z = 1.5m$$
$$\rho_L = 1000 \text{ kg/m}^3$$
$$\mu_L = 0.01\rho_L^{0.5} = 0.01.\sqrt{(1000)} = 0.316\text{mNs/m}^2$$
$$\phi = 2.6$$
$$M = 18 \text{ kg/kmol}$$
$$T = 303 \text{ K}$$

$$D_L = \frac{1.173x10^{-13}\,(\phi.M)^{0.5}\,T}{\mu_S.V_M^{0.6}}$$

$$= \frac{1.173x10^{-13}\,(2.6x18)^{0.5}\,x303}{0.316x(0.034)^{0.6}} = 5.85x10^{-9}, \frac{m^2}{s}$$

$$Sc_L = \frac{\mu_L}{\rho_L D} = \frac{3.16x10^{-4}}{1000x5.85x10^{-9}}, \frac{kg}{m.s}.\frac{m^3}{kg}.\frac{s}{m^2} = 54$$

Then

$$H_L = 0.305\Phi_H \left(Sc_L\right)^{0.5} K_3 \cdot \left(\frac{Z}{3.05}\right)^{0.15}$$

$$= 0.305 x\, 0.064 x \left(54\right)^{0.5} x\, 0.85 x \left(\frac{1.5}{3.05}\right)^{0.15} = 0.110\,m. \quad Ans.$$

Example 4.10

According to Onda *et al* (J. Chem. Eng., Japan, 1968), the effective area, a_W, for wetted packing above 15 mm size, is given by

$$\frac{a_W}{a} = 1 - \exp\left[-1.45\left(\frac{\sigma_C}{\sigma_L}\right)^{0.75}\left(\frac{L}{a\,\mu_L}\right)^{0.1}\left(\frac{L^2 a}{\rho_L^2 g}\right)^{-0.05}\left(\frac{L^2}{\rho_L \sigma_L a}\right)^{0.2}\right] \quad (1)$$

The gas film coefficient, k_G, is given by

$$\frac{k_G}{a}\cdot\frac{RT}{D_V} = 5.23\left(\frac{V_W}{a\,\mu_V}\right)^{0.7}\left(\frac{\mu_V}{\rho_V D_V}\right)^{1/3}\left(a\,d_p\right)^{-2.0} \quad (2)$$

and the height of a transfer unit (HTU), H_G, by

$$H_G = \frac{G_M}{k_G\, a_W\, P} \quad (3)$$

If

σ_C	$= 61 \times 10^{-3}$ N/m
σ_L	$= 72 \times 10^{-3}$ N/m
R	$= 0.08206$ atm.m³/kmol.K
T	$= 300$ K
L	$= 0.65$ kg/m².s
μ_L	$= 8.5 \times 10^{-6}$ Ns/m²
μ_V	$= 1.5 \times 10^{-5}$ Ns/m²
ρ_L	$= 998$ kg/m³
ρ_V	$= 1.8$ kg/m³
a	$= 240$ m²
g	$= 9.81$ m/s²
P	$= 1$ atm.
d_P	$= 19.05 \times 10^3$ m
G_M	$=$ molar gas rate, kmol/s. m²
k_G	$=$ mass transfer coefficient in the gas film, kmol/s. m².atm
V_W	$= 0.95$ kg/ m².s

determine the height of a transfer unit for this case.

Answer

This is a problem in which the ability to substitute and evaluate correlations, correctly, is demonstrated. To evaluate

$$k_G = 5.23 \frac{a D_V}{RT} \cdot \left(\frac{V_W}{a\,\mu_V}\right)^{0.7} \left(\frac{\mu_V}{\rho_V D_V}\right)^{1/3} \left(a d_p\right)^{-2.0} \qquad from \quad (2)$$

$$\frac{a D_V}{RT} = \frac{240 \times 1.77 \times 10^{-9}}{0.08206 \times 300} \cdot \frac{m^2}{m^3} \cdot \frac{m^2}{s} \cdot \frac{kmol.K}{atm.m^3} \cdot \frac{1}{K} = 1.726 \times 10^{-8} \frac{kmol}{m^2.s.atm} \quad (4)$$

$$\left(\frac{V_W}{a\,\mu_V}\right)^{0.7} = \left(\frac{0.95}{240 \times 1.5 \times 10^{-5}}, \frac{kg}{m^2.s} \frac{m^3}{m^2} \frac{m.s}{kg}\right)^{0.7} = 49.4, \text{dim}ensionless \quad (5)$$

$$\left(\frac{\mu_V}{\rho_V D_V}\right)^{1/3} = \left(\frac{1.5 \times 10^{-5}}{1.8 \times 1.77 \times 10^{-9}}, \frac{kg}{m.s} \cdot \frac{m^3}{kg} \cdot \frac{s}{m^2}\right)^{1/3} = 16.76, \text{dim}ensionless \quad (6)$$

$$\left(a d_p\right)^{-2.0} = \left(240 \times 19.05 \times 10^{-3}, \frac{m^2}{m^3} \cdot \frac{m}{1}\right)^{-2.0} = 0.048, \text{dim}ensionless \quad (7)$$

Thus

$$k_G = 5.23 \times 1.726 \times 10^{-8} \times 49.54 \times 16.76 \times 0.048, \frac{kmol}{m^2.s.atm}$$

$$= 3.598 \times 10^{-8} \frac{kmol}{m^2.s.atm} \quad (8)$$

To evaluate

$$\frac{a_W}{a} = 1 - \exp\left[-1.45\left(\frac{\sigma_C}{\sigma_L}\right)^{0.75}\left(\frac{L}{a\,\mu_L}\right)^{0.1}\left(\frac{L^2 a}{\rho_L^2 g}\right)^{-0.05}\left(\frac{L^2}{\rho_L \sigma_L a}\right)^{0.2}\right]$$

$$from \quad (1)$$

$$\left(\frac{\sigma_C}{\sigma_L}\right)^{0.75} = \left(\frac{61}{72}\right)^{0.75} = 0.883, \text{dim}ensionless \quad (9)$$

$$\left(\frac{L}{a\,\mu_L}\right)^{0.1} = \left(\frac{0.65}{240 \times 8.5 \times 10^{-6}}, \frac{kg}{m^2.s} \frac{m^3}{m^2} \frac{m.s}{kg}\right)^{0.1}$$

$$= 1.78, \text{dim} \, ensionless \tag{10}$$

$$\left(\frac{L^2 \, a}{\rho_L^2 \, g} \right)^{-0.05} = \left(\frac{(0.65)^2 \, x \, 240}{(998)^2 \, x \, 9.81}, \frac{kg^2}{m^4 . s^2} \cdot \frac{m^2}{m^3} \cdot \frac{m^6}{kg^2} \cdot \frac{s^2}{m} \right)^{-0.05}$$

$$= 1.775, \text{dim} \, ensionless \tag{11}$$

$$\left(\frac{L^2}{\rho_L \, \sigma_L \, a} \right)^{0.2} = \left(\frac{(0.65)^2}{998 x \, 72 x 10^{-3} \, x \, 240}, \frac{kg^2}{m^4 . s^2} \cdot \frac{m^3}{kg} \cdot \frac{m}{N} \cdot \frac{m^3}{m^2} \right)^{0.2}$$

$$= 0.1196, \text{dim} \, ensionless \tag{12}$$

Thus

$$\frac{a_W}{a} = 1 - \exp\left[-1.45 x \, 0.883 x 1.78 x 1.775 x 0.1196\right] = 0.384$$

from which

$$a_W = 0.384 x \, 240 \, m^2 / m^3 = 92.16 \, m^2 / m^3 \tag{13}$$

Finally

$$H_G = \frac{G_M}{k_G \, a_W \, P} = \frac{0.95}{44 x \, 3.598 x 10^{-6} \, x \, 92.16 x 1},$$

$$\left(\frac{kg}{s.m^2} \cdot \frac{kmol}{kg} \cdot \frac{m^2 . s \, . atm}{kmol} \cdot \frac{m^3}{m^2} \cdot \frac{1}{atm} \right) = 65.11 \, m. \quad Ans.$$

References for Chapter Four

1. Coulson J. M., Richardson J. F., Backhurst J. R. and Harker J. H., *Chemical Engineering*, Vol. II, 3rd. Edn. (SI Units); Pergamon Press, Oxford, UK, 1980

2. Coulson J. M., Richardson J. F. and Sinnott R. K., *Chemical Engineering*, Vol. 6 (SI Units); *Design*; Pergamon Press, Oxford, UK, 1983

3   Onda K, Takeuchi H and Okumoto Y (1968); *Mass Transfer Coefficients between Gas and Liquid Phases in Packed Columns*, J. Chem. Eng., **I**; p56; in Coulson J. M., Richardson J. F. and Sinnott R. K (opus cit.).

APPENDIX I
General Equations for Momentum, Heat and Mass Transfer in Cartesian Co-ordinates

1. Equations of Motion for an Incompressible Newtonian Fluid

x-component

$$\frac{\partial u}{\partial t} + u\frac{\partial u}{\partial x} + v\frac{\partial u}{\partial y} + w\frac{\partial u}{\partial z} = g_x - \frac{1}{\rho}\frac{\partial P}{\partial x} + \frac{\mu}{\rho}\left(\frac{\partial^2 u}{\partial x^2} + \frac{\partial^2 u}{\partial y^2} + \frac{\partial^2 u}{\partial z^2}\right)$$

y - component

$$\frac{\partial v}{\partial t} + u\frac{\partial v}{\partial x} + v\frac{\partial v}{\partial y} + w\frac{\partial v}{\partial z} = g_y - \frac{1}{\rho}\frac{\partial P}{\partial y} + \frac{\mu}{\rho}\left(\frac{\partial^2 v}{\partial x^2} + \frac{\partial^2 v}{\partial y^2} + \frac{\partial^2 v}{\partial z^2}\right)$$

z – component

$$\frac{\partial w}{\partial t} + u\frac{\partial w}{\partial x} + v\frac{\partial w}{\partial y} + w\frac{\partial w}{\partial z} = g_z - \frac{1}{\rho}\frac{\partial P}{\partial z} + \frac{\mu}{\rho}\left(\frac{\partial^2 w}{\partial x^2} + \frac{\partial^2 w}{\partial y^2} + \frac{\partial^2 w}{\partial z^2}\right)$$

2. Equation of Energy Transfer for an Incompressible Newtonian Fluid

$$\frac{\partial T}{\partial t} + u\frac{\partial T}{\partial x} + v\frac{\partial T}{\partial y} + w\frac{\partial T}{\partial z} = \frac{k}{\rho Cp}\cdot\left(\frac{\partial^2 T}{\partial x^2} + \frac{\partial^2 T}{\partial y^2} + \frac{\partial^2 T}{\partial z^2}\right) + \frac{Q_V}{\rho Cp}$$

3. Equation of Mass Transfer for an Incompressible Newtonian Fluid

$$\frac{\partial C_A}{\partial t} + u\frac{\partial C_A}{\partial x} + v\frac{\partial C_A}{\partial y} + w\frac{\partial C_A}{\partial z} = D_{AB}\cdot\left(\frac{\partial^2 C_A}{\partial x^2} + \frac{\partial^2 C_A}{\partial y^2} + \frac{\partial^2 C_A}{\partial z^2}\right) + R_{A_V}$$

APPENDIX II
General Equations for Momentum, Heat and Mass Transfer in Cylindrical Polar Co-ordinates

1. Equations of Motion for an Incompressible Newtonian Fluid

r-component

$$\frac{\partial V_r}{\partial t} + V_r \frac{\partial V_r}{\partial r} + \frac{V_\theta}{r} \frac{\partial V_r}{\partial \theta} - \frac{V_\theta^2}{r} + V_z \frac{\partial V_r}{\partial z} = g_r - \frac{1}{\rho} \frac{\partial P}{\partial r}$$

$$+ \frac{\mu}{\rho} \left[\frac{\partial}{\partial r} \left(\frac{1}{r} \frac{\partial (r V_r)}{\partial r} \right) + \frac{1}{r^2} \frac{\partial^2 V_r}{\partial \theta^2} - \frac{2}{r^2} \frac{\partial V_\theta}{\partial \theta} + \frac{\partial^2 V_r}{\partial z^2} \right]$$

θ - component

$$\frac{\partial V_\theta}{\partial t} + V_r \frac{\partial V_\theta}{\partial r} + \frac{V_\theta}{r} \frac{\partial V_\theta}{\partial \theta} + \frac{V_r V_\theta}{r} + V_z \frac{\partial V_\theta}{\partial z} = g_\theta - \frac{1}{\rho r} \frac{\partial P}{\partial \theta}$$

$$+ \frac{\mu}{\rho} \left[\frac{\partial}{\partial r} \left(\frac{1}{r} \frac{\partial (r V_\theta)}{\partial r} \right) + \frac{1}{r^2} \frac{\partial^2 V_\theta}{\partial \theta^2} - \frac{2}{r^2} \frac{\partial V_r}{\partial \theta} + \frac{\partial^2 V_\theta}{\partial z^2} \right]$$

z – component

$$\frac{\partial V_z}{\partial t} + \frac{V_r}{r} \frac{\partial V_z}{\partial r} + \frac{V_\theta}{r} \frac{\partial V_z}{\partial \theta} + V_z \frac{\partial V_z}{\partial z} = g_z - \frac{1}{\rho} \frac{\partial P}{\partial z}$$

$$+ \frac{\mu}{\rho} \left[\frac{1}{r} \frac{\partial}{\partial r} \left(r \frac{\partial V_z}{\partial r} \right) + \frac{1}{r^2} \frac{\partial^2 V_z}{\partial \theta^2} + \frac{\partial^2 V_z}{\partial z^2} \right]$$

2. Heat Energy Transfer Equations for an Incompressible Newtonian Fluid

$$\frac{\partial T}{\partial t} + V_r \frac{\partial T}{\partial r} + \frac{V_\theta}{r} \frac{\partial T}{\partial \theta} + V_z \frac{\partial T}{\partial z}$$

$$= \frac{k}{\rho C p} \left[\frac{1}{r} \frac{\partial}{\partial r} \left(\frac{r \partial T}{\partial r} \right) + \frac{1}{r^2} \frac{\partial^2 T}{\partial \theta^2} + \frac{\partial^2 T}{\partial z^2} \right] + \frac{Q_V}{\rho C p}$$

3. Mass Transfer Equations for an Incompressible Newtonian Fluid

$$\frac{\partial C_A}{\partial t} + V_r \frac{\partial C_A}{\partial r} + \frac{V_\theta}{r} \frac{\partial C_A}{\partial \theta} + V_z \frac{\partial C_A}{\partial z}$$

$$= D_{AB} \left[\frac{1}{r} \frac{\partial}{\partial r} \left(r \frac{\partial C_A}{\partial r} \right) + \frac{1}{r^2} \frac{\partial^2 C_A}{\partial \theta^2} + \frac{\partial^2 C_A}{\partial z^2} \right] + R_{A_V}$$

APPENDIX III
General Equations for Momentum, Heat and Mass Transfer in Spherical Co-ordinates

1. Equations of Motion for an Incompressible Newtonian Fluid

r-component

$$\frac{\partial V_r}{\partial t} + V_r \frac{\partial V_r}{\partial r} + \frac{V_\theta}{r} \frac{\partial V_r}{\partial \theta} - \frac{V_\theta^2 + V_\phi^2}{r} + \frac{V_\phi}{r \sin\theta} \frac{\partial V_r}{\partial \phi} = g_r - \frac{1}{\rho} \frac{\partial P}{\partial r}$$

$$+ \frac{\mu}{\rho} \left[\nabla^2 V_r - \frac{2}{r^2} V_r - \frac{2}{r^2} \frac{\partial V_\theta}{\partial \theta} - \frac{2 V_\theta \cot\theta}{r^2} - \frac{2}{r^2 \sin\theta} \frac{\partial V_\phi}{\partial \phi} \right]$$

θ - component

$$\frac{\partial V_\theta}{\partial t} + V_r \frac{\partial V_\theta}{\partial r} + \frac{V_\theta}{r} \frac{\partial V_\theta}{\partial \theta} + \frac{V_r V_\theta}{r} + \frac{V_\phi}{r \sin\theta} \frac{\partial V_\theta}{\partial \phi} - \frac{V_\phi^2 \cot\theta}{r} = g_\theta - \frac{1}{\rho r} \frac{\partial P}{\partial \theta}$$

$$+ \frac{\mu}{\rho} \left[\nabla^2 V_\theta + \frac{2}{r^2} \frac{\partial V_r}{\partial \theta} - \frac{V_\theta}{r^2 \sin^2\theta} - \frac{2\cos\theta}{r^2 \sin^2\theta} \frac{\partial V_\phi}{\partial \phi} \right]$$

ϕ - component

$$\frac{\partial V_\phi}{\partial t} + V_r \frac{\partial V_\phi}{\partial r} + \frac{V_\theta}{r} \frac{\partial V_\phi}{\partial \theta} + \frac{V_\phi}{r \sin\theta} \frac{\partial V_\theta}{\partial \phi} + \frac{V_\phi V_r}{r} + \frac{V_\phi V_\theta}{r} \cot\theta$$

$$= g_\phi - \frac{1}{\rho r \sin\theta \partial\phi} \frac{\partial P}{} + \frac{\mu}{\rho} \left[\nabla^2 V_\phi - \frac{V_\phi}{r^2 \sin^2\theta} + \frac{2}{r^2 \sin\theta} \frac{\partial V_r}{\partial \phi} + \frac{2\cos\theta}{r^2 \sin^2\theta} \frac{\partial V_\theta}{\partial \phi} \right]$$

where

$$\nabla^2 = \frac{1}{r^2} \frac{\partial}{\partial r} \left(r^2 \frac{\partial}{\partial r} \right) + \frac{1}{r^2 \sin\theta} \frac{\partial}{\partial \theta} \left(\sin\theta \frac{\partial}{\partial \theta} \right) + \frac{1}{r^2 \sin^2\theta} \frac{\partial^2}{\partial \phi^2}$$

2. Equation of Energy Transfer for an Incompressible Newtonian Fluid

$$\frac{\partial T}{\partial t} + V_r \frac{\partial T}{\partial r} + \frac{V_\theta}{r} \frac{\partial T}{\partial \theta} + \frac{V_\phi}{r \sin\theta} \frac{\partial T}{\partial \phi} = + \frac{Q_V}{\rho Cp}$$

$$+ \frac{k}{\rho Cp} \left[\frac{1}{r^2} \frac{\partial}{\partial r} \left(r^2 \frac{\partial T}{\partial r} \right) + \frac{1}{r^2 \sin\theta} \frac{\partial}{\partial \theta} \left(\sin\theta \frac{\partial T}{\partial \theta} \right) + \frac{1}{r^2 \sin^2\theta} \frac{\partial^2 T}{\partial \phi^2} \right]$$

3. **Equation of Mass Transfer for an Incompressible Newtonian Fluid**

$$\frac{\partial C_A}{\partial t} + V_r \frac{\partial C_A}{\partial r} + \frac{V_\theta}{r} \frac{\partial C_A}{\partial \theta} + \frac{V_\phi}{r \sin \theta} \frac{\partial C_A}{\partial \phi} = R_{A_V}$$

$$+ D_{AB} \left[\frac{1}{r^2} \frac{\partial}{\partial r} \left(r^2 \frac{\partial C_A}{\partial r} \right) + \frac{1}{r^2 \sin \theta} \frac{\partial}{\partial \theta} \left(\sin \theta \frac{\partial C_A}{\partial \theta} \right) + \frac{1}{r^2 \sin^2 \theta} \frac{\partial^2 C_A}{\partial \phi^2} \right]$$

APPENDIX IV
Densities Of Various Materials

Density of water (at 1 atm)		Density of air (at 1 atm)	
Temp (°C)	Density (kg/m³)	Temp (°C)	Density (kg/m³)
100	958.4	−25	1.423
80	971.8	−20	1.395
60	983.2	−15	1.368
40	992.2	−10	1.342
30	995.6502	−5	1.316
25	997.0479	0	1.293
22	997.7735	5	1.269
20	998.2071	10	1.247
15	999.1026	15	1.225
10	999.7026	20	1.204
4	999.9720	25	1.184
0	999.8395	30	1.164
−10	998.117	35	1.146
−20	993.547		
−30	983.854		

The values below 0 °C refer to super cooled water.

Density of Solutions

The density of a solution, ρ, is the sum of mass concentrations of the components of that solution. The mass concentration of a given component, ρ_i, in a solution can be called the partial density of that component.

$$\rho = \sum_i \rho_i$$

Density of Liquids

Liquid	Density, kg/m^3	Liquid	Density, kg/m^3
Acetic acid, liquid	1057	Oil, petroleum crude	849
Acetone	785	Oil, sperm whale	913
Alcohol, ethyl	897	Oil, turpentine	865
Alcohol, methyl	785	Linseed oil	929
Asphalt, liquid	1041	Maple syrup	1362
Aviation fuel (jp-4)	785	Methanol	785
Carbon tetrachloride		Mineral oil	913
Citric acid	881	Mineral spirits	785
Corn, sugar, liquid	1410	Nitric acid	1506
Cottonseed oil	929	Octane	721
Diesel fuel	833	Oil, olive	913
Ethanol	897	Oil, transformer	881
Ethyl ether	705	Petroleum oil	817
Ethylene glycol	1121	Sea water	1025
Gasoline	721	Sulfuric acid	1794
Glycerine	1249	Transmission oil	865
Hydrochloric acid	1202	Water	993
Kerosene	817		

Reference: (*en.wikipedia.org/wiki/Density*)

APPENDIX V
Viscosities Of Various Materials

Viscosity of air

The viscosity of air depends mostly on the temperature. At 15.0 °C, the viscosity of air is 1.78×10^{-5} kg/(m·s), 17.8 µPa.s or 1.78×10^{-5} Pa.s..

	Viscosity of selected gases at 100 kPa, [µPa·s]		
Gas	at 0 °C		at 27 °C
Air	17.4	18.27 (18 C)	18.6
Ammonia		9.82 (19 C)	10.08
Argon			22.9
Carbon dioxide		14.80 (19 C)	15.15
Carbon monoxide		17.20 (15 C)	17.78
Ethane			9.50
Hydrogen	8.4	8.76 (21 C)	8.90
Helium			20.0
Methane			11.20
Nitrogen			17.81
Oxygen		20.18 (19 C)	20.64
Sulphur dioxide		12.54 (21 C)	12.86
Xenon	21.2		23.20

References

1. http://en.wikipedia.org/wiki/Viscosity
2. Chemical Rubber Company (CRC). 1984. CRC Handbook of Chemistry and Physics. Weast, Robert C., editor. 65th edition. CRC Press, Inc. Boca Raton, Florida. USA.

Viscosity of water

T, °C	Viscosity, mPa·s	T, °C	Viscosity, mPa·s	T, °C	Viscosity, mPa·s
10	1.308	40	0.6531	80	0.3550
20	1.002	50	0.5471	90	0.3150
25	0.890	60	0.4668	100	0.2822
30	0.7978	70	0.4044		

Viscosity of Newtonian liquids at 25 °C

Liquid	Viscosity, (Pa·s)	Viscosity, (cP=mPa.s)
Acetone	3.06×10^{-4}	0.306
Benzene	6.04×10^{-4}	0.604
castor oil	0.985	985
corn syrup	1.3806	1380.6
Ethanol	1.074×10^{-3}	1.074
ethylene glycol	1.61×10^{-2}	16.1
glycerol	1.49 (at 20 °C)	1490
Fuel oil, HFO-380	2.022	2022
Mercury	1.526×10^{-3}	1.526
Methanol	5.44×10^{-4}	0.544
Nitrobenzene	1.863×10^{-3}	1.863
liquid nitrogen @ 77K	1.58×10^{-4}	0.158
propanol	1.945×10^{-3}	1.945
olive oil	.081	81
sulphuric acid	2.42×10^{-2}	24.2
Water	8.94×10^{-4}	0.894

Viscosity of non-Newtonian liquids

Fluid	Viscosity (Pa·s)	Viscosity (cP)
Honey	2–10	2,000–10,000
Molasses	5–10	5,000–10,000
molten glass	10–1,000	10,000–1,000,000
chocolate syrup	10–25	10,000–25,000
molten chocolate*	45–130	45,000–130,000
ketchup*	50–100	50,000–100,000
peanut butter*	c. 250	c. 250,000
shortening*	c. 250	250,000

* These materials are highly non-Newtonian

Viscosity of blends of liquids

The viscosity of the blend of two or more liquids can be estimated using the Refutas equation. The calculation is carried out in three steps.

Step 1: calculate the Viscosity Blending Number (VBN) (also called the Viscosity Blending Index) of each component of the blend:

$$VBN = 14.534 \, x \ln\left[\ln\left(v + 0.8\right)\right] + 10.975 \tag{1}$$

where v is the kinematic viscosity in centistokes (cSt). It is important that the kinematic viscosity of each component of the blend be obtained at the same temperature.

Step 2: calculate the VBN of the blend using equation (2)

$$VBN_{blend} = \sum_{i=1}^{N}\left(x_i \, x \, VBN_i\right) \tag{2}$$

where x_i is the mass fraction of component i of the blend.

Step 3: determine the kinematic viscosity of the blend by solving equation (1) for v:

$$v = \exp\left[\exp\left(\frac{VBN_{blend} - 10.975}{14.534}\right)\right] - 0.8 \tag{3}$$

where VBN_{Blend} is the viscosity blending number of the blend.

APPENDIX VI
Diffusion Coefficients

Table 1.
Oxygen Diffusion Coefficients of Binary Gas Pairs at Atmospheric Pressure

Binary Pair	Temp [°C]	Diffusion Coefficient [cm²/s]
Oxygen - Carbon Dioxide	20	0.153
	60	0.193
Oxygen - Water Vapour	20	0.240
	60	0.339
Oxygen – Nitrogen	20	0.219
	60	0.274

Reference: www.wag.caltech.edu/home/jang/genchem/diffus.htm

Table 2.
Effect of Temperature and Gas Mixture on Oxygen Diffusion Coefficients

		Oxygen Diffusion Coefficient [cm²/s]	
Temperature	Relative Humidity	at 2% O_2	at 15% O_2
20°C	50%	0.203	0.214
20°C	100%	0.203	0.214
60°C	50%	0.259	0.273
60°C	100%	0.264	0.278

Reference: Tom Richard; Calculating the Oxygen Diffusion Coefficient in Air Tom Richard, Cornell Waste Management Institute, Cornell University, Ithaca, NY. cwmi@cornell.edu

Table 3: Diffusion Coefficients, D, at 37 C

Substance	M. Wt	$10^9 D$, m²/s	Substance	M. Wt	$10^9 D$, m²/s
H_2	2	5.40	KCl	75	2.00
H_2O	18	2.31	Glycerol	92	8.80
O_2	32	2.00	$CaCl_2$	111	1.20
Methanol	32	1.50	Glucose	180	71.00
HCl	36.5	3.60	Citric acid	192	69.00
CO_2	44	1.90	Sucrose	342	54.00
NaCl	58.5	1.50			

Extracted from a Table by nanomedicine.com, 2003; Volume IIA, Biocompatibility, Landes Bioscience, Georgetown, Texas.